Bioactive Molecules in Plant Defense

Mostafa Abdelrahman • Sudisha Jogaiah

Bioactive Molecules in Plant Defense

Saponins

Springer

Mostafa Abdelrahman
Botany Department, Faculty of Science
Aswan University
Aswan, Egypt

Sudisha Jogaiah
Department of Biotechnology
Karnatak University
Dharwad, India

ISBN 978-3-030-61148-4 ISBN 978-3-030-61149-1 (eBook)
https://doi.org/10.1007/978-3-030-61149-1

This Springer imprint is published by the registered company Springer Nature Switzerland AG.
The registered company address is: Gewerbestrasse 11, 6330 Cham, Switzerland

Preface

Saponins are a large group of bioactive compounds which are present in most of the medicinal and crop plant species. Saponin compounds are characterized by antimicrobial and pharmaceutical properties; thus, they are of great interest for both plant disease management and pharmaceutical industries. Saponin compounds can be classified based on their chemical structure into two main classes including triterpene saponins and steroidal saponins. Although the basic biosynthesis pathway through mevalonate kinase and plastidial methylerythritol phosphate pathway are involved in the upstream biosynthesis of saponin precursors, the downstream biosynthesis pathway of saponin remains elusive. The molecular diversity of oxidosqualene cyclase (OSC), which catalyzes the first diversifying step in steroidal saponin and triterpenoid biosynthesis, results in a diverse array of saponin substrates. After the basic skeleton is constructed by OSCs, the skeleton is tailored into a hydrophobic aglycone by cytochrome P450 monooxygenase, followed by further modification through glycosylation process.

In this book, we will summarize and discuss the different aspects of saponin compounds and their roles in plant defense, including (I) general introduction of the saponin compounds, (II) production of plant bioactive triterpenoid and steroidal saponins, (III) metabolic and functional diversity of saponins, (IV) saponins versus plant fungal pathogens, (V) saponin-detoxifying enzymes, (VI) isolation and characterization of triterpenoid and steroidal saponins, and (VII) method of estimation in biological sample and finally genetic engineering of saponin and the target genes to improve yields. The discovery of biosynthesis, transcriptional factor(s), and transporter genes involved in saponin biosynthesis is a crucial leap for stable production and further promising applications of the saponin compounds in agrochemical or pharmaceutical industries.

Aswan, Egypt — Mostafa Abdelrahman
Dharwad, India — Sudisha Jogaiah

Contents

Introduction

1

1.1 Introduction

There are approximately 450,000 plant species exist on earth, and one third of these plants are under the risk of extinction (Pimm and Joppa 2015). The current estimated total number of plant-produced metabolites within a given plant species are greater than 10,000, however, at present, it has been projected that with currently available metabolome techniques, only less than 20% of these metabolites can be analyzed (Lei et al. 2011; Abdelrahman et al. 2018). During life span, human has frequently used plant derived natural products as traditional medicines for millennia. However, the full potential of these plant derived natural products remains to be exploited, because they are difficult to synthesis in vitro, exist in very low amounts in a given plant species and/or produced by rare plant species and thus cannot be utilized for the large scale production. Generally speaking plants synthesize a diverse array of primary and secondary metabolites, which have different structures and vary greatly in their richness (Arbona et al. 2013; Hong et al. 2016). For instance, primary metabolites are crucial for plant growth and development, whereas secondary metabolites have more explicit functions; and both types of metabolites have major roles for plant responses to a specific stress (Fujii et al. 2015; Abdelrahman et al. 2019).

Saponin compounds are classified as secondary metabolites with remarkable chemical structure, and distinguishable biological activities (Mostafa et al. 2013; Abdelrahman et al. 2014). Plants frequently produce saponins as part of their common cycle of growth and development as basic chemical barriers for defense mechanisms against pathogenic fungi and insects (Abdelrahman et al. 2017). Since many saponin compounds display effective antifungal activity and are usually found in relatively high amounts in healthy plants, these saponin compounds have been considered as determinants of a plant's resistance to pathogenic fungi (Osbourn 1996) (Fig. 1.1).

Saponins exhibit a wide range of biological properties, such as emulsifying and foaming, medicinal and pharmacological properties, sweetness and bitterness, as

M. Abdelrahman, S. Jogaiah, *Bioactive Molecules in Plant Defense*,
https://doi.org/10.1007/978-3-030-61149-1_1

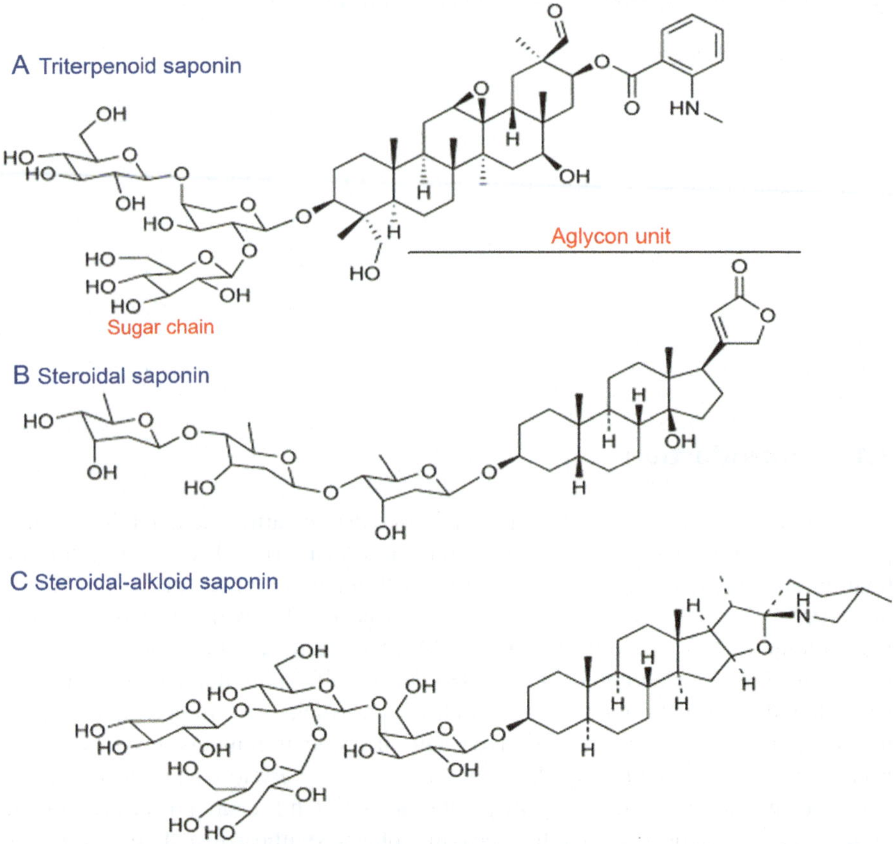

Fig. 1.1 Classification of triterpene (**a**), steroidal (**b**) and steroidal-alkaloid (**c**) saponin compounds based on their aglycon unit structure. Saponins can be classified based on their chemical structure into three main groups namely triterpenoid, steroid, or steroidal glycoalkaloid, according to the structure of the aglycone or sapogenin unit, which represent the core of the structure. Steroidal saponins are also sub-classified into furostanol and spriostanol as two main saponin compounds usually detected. Triterpenoid saponins are found mostly in dicotyledonous plants, however they can be also found in some monocots, whereas most of the steroid saponin compounds were isolated mainly from monocots (Abe et al. 1993; Connolly and Hill 1991; Hostettmann and Marston 1995; Sparg et al. 2004). Early study by Vincken et al. (2007), provided further detailed classification of the saponin compounds and they grouped the saponins based on their structure into 11 major classes of saponin compounds that can be identified and detected in different plant species, including lupanes, tirucallanes, dammaranes, hopanes, oleananes, ursanes, taraxasteranes, cycloartanes, cucurbitanes, lanostanes, and steroids. Furthermore, the ursanes, lupanes, dammaranes, hopanes and oleananes and steroids can be further subdivided into 16 subclasses, because their carbon bones are subjected to homologation, fragmentation and degradation (Xu et al. 2004; Vincken et al. 2007)

well as antifungal, insecticidal, and molluscicidal properties (Oda et al. 2000; Kitagawa 2002; Sparg et al. 2004; Heng et al. 2006). Saponin compounds are involved in a wide range of applications, especially in cosmetic, pharmaceutical and beverage industries (Price et al. 1987; Petit et al. 1995; Uematsu et al. 2000;

Sparg et al. 2004; Abdelrahman et al. 2017). Saponins exhibit detergent characteristics due to the nature of sugar chain moieties and aglycon or saponin unit. For instance, the sugar moieties are water-soluble and thus saponin compounds can be soluble in water, whereas the sapogenin or aglycon unit is fat-soluble (Tan et al. 1999; Connolly and Hill 2000). The stability of the biological activity of saponin compounds to heat processing and normal cooking has been reported (Vincken et al. 2007). Several chromatographic techniques have been used to isolate pure saponin compounds from plant materials, which mostly include extraction step using a weak or nonpolar solvents such as petroleum ether or chloroform to remove lipids, followed by polar solvent extraction such as methanol or ethanol to obtain the pure saponin extract, then the pure saponin extracts being subjected to various purification process using column chromatography (Mostafa et al. 2013; Abdelrahman et al. 2014, 2017). After isolation of pure saponin compounds, different structure elucidation techniques are applied to identify the chemical structure of the saponin. However, the analysis of saponin compounds is complex due to the long process of isolation, separation, identification and quantification steps. Thus, in this book, the recent developments with regards to the roles of the saponin compounds in plant defense, and the saponin-pathogen relationship are introduced, and finally we would like to introduce some of the techniques used for saponin isolation and quantification which can be a guide for students as well as high professional labs.

References

Abdelrahman M, Hirata S, Ito S, Yamauchi N, Shigyo M (2014) Compartmentation and localization of bioactive metabolites in different organs of *Allium roylei*. Biosci Biotechnol Biochem 78:1112–1122

Abdelrahman M et al (2017) RNA-sequencing-based transcriptome and biochemical analyses of steroidal saponin pathway in a complete set of *Allium fistulosum—A. cepa* monosomic addition lines. PLoS One, 12:e0181784

Abdelrahman M, Burritt DJ, Tran LP (2018) The use of metabolomic quantitative trait locus mapping and osmotic adjustment traits for the improvement of crop yields under environmental stresses. Semin Cell Dev Biol 83:86–94

Abdelrahman M, Hirata S, Sawada Y, Hirai MY, Sato S, Hirakawa H, Mine Y, Tanaka K, Shigyo M (2019) Widely targeted metabolome and transcriptome landscapes of Allium fistulosum–A. cepa chromosome addition lines revealed a flavonoid hot spot on chromosome 5A. Sci Rep 9:3541

Abe I, Rohmer M, Prestwich GC (1993) Enzymatic cyclization of squalene and oxidosqualene to sterols and triterpenes. Chem Rev 93:2189–2206

Arbona V, Manzi M, de Ollas C, Gómez-Cadenas A (2013) Metabolomics as a tool to investigate abiotic stress tolerance in plants. Int J Mol Sci 14:4885–4911

Connolly JD, Hill RA (1991) Dictionary of Terpenoids, vol I, xliii–xlvii, II. Chapman & Hall, New York, pp 1121–1415

Connolly JD, Hill RA (2000) Triterpenoids. Nat Prod Rep 17:463–482

Fujii T, Matsuda S, Tejedor ML, Esaki T, Sakane I, Mizuno H, Tsuyama N, Masujima T (2015) Direct metabolomics for plant cells by live single-cell mass spectrometry. Nat Protoc 10:1445–1456

Heng L, Vincken JP, van Koningsveld GA, Legger L, Gruppen H, van Boekel MAJS, Roozen JP, Voragen AGJ (2006) Bitterness of saponins and their content in dry peas. J Sci Food Agric 86:1225–1231

Hong J, Yang L, Zhang D, Shi J (2016) Plant metabolomics: an indispensable systembiology tool for plant science. Int J Mol Sci 17:E767

Hostettmann K, Marston A (1995) Saponins. Cambridge University Press, Cambridge. https://doi.org/10.1017/CBO9780511565113

Kitagawa I (2002) Licorice root. A natural sweetener and an important ingredient in Chinese medicine. Pure Appl Chem 74:1189–1198

Lei Z, Huhman DV, Sumner LW (2011) Mass spectrometry strategies in metabolomics. J Biol Chem 286(29)

Mostafa A, Sudisha J, El-Sayed M, Ito S-I, Ikeda T, Yamauchi N, Shigyo M (2013) Aginoside saponin, a potent antifungal compound, and secondary metabolite analyses from *Allium nigrum* L. Phytochem Lett 6:274–280

Oda K, Matsuda H, Murakami T, Katayama S, Ohgitani T, Yoshikawa M (2000) Adjuvant and haemolytic activities of 47 saponins derived from medicinal and food plants. Biol Chem 381:67–74

Osbourn AE (1996) Preformed antimicrobial compounds and plant defense against fungal attack. Plant Cell 8:1821–1831

Petit PR, Sauvaire YD, Hillaire-Buys DM, Leconte OM, Baissac YG, Posin GR, Ribes GR (1995) Steroid saponins from fenugreek seeds: extraction, purification, and pharmacological investigation on feeding behaviour and plasma cholesterol. Steroids 60:674–680

Pimm SL, Joppa LN (2015) How many plant species are there, where are they, and at what rate are they going extinct? Ann Mo Bot Gard 100:170–176

Price KR, Johnson IT, Fenwick GR (1987) The chemistry and biological significance of saponins in foods and feedstuffs. Crit Rev Food Sci Nutr 26:127–135

Sparg SG, Light ME, van Staden J (2004) Biological activities and distribution of plant saponins. J Ethnopharmacol 94:219–243

Tan N, Zhou J, Zhao S (1999) Advances in structural elucidation of glucuronide oleanane-type triterpene carboxylic acid 3,28-O-bisdesmosides (1962–1997). Phytochemistry 52:153–192

Uematsu Y, Hirata K, Saito K (2000) Spectrophotometric determination of saponin in Yucca extract used as food additive. J AOAC Int 83:1451–1454

Vincken J-P, Heng L, Groot A, Gruppen H (2007) Saponins, classification and occurrence in the plant kingdom. Phytochemistry 68:275–297

Xu R, Fazio GC, Matsuda PT (2004) On the origins of triterpenoid skeletal diversity. Phytochemistry 65:261–291

Production of Plant Bioactive Triterpenoid and Steroidal Saponins

2

Abstract

Saponins are triterpenoid or steroid glycosides that play various biological activities in different plant species. The broad-spread existence of saponin in several plant species, and the potential for saponin pharmaceutical applications have resulted in the extraction and identification of many saponin compounds in various plant species. Although an extensive efforts have been invested in the extraction, isolation and chemical structure identification of saponin compounds, which are important to extend our knowledge of saponin structures, recent progress has been given to the biosynthesis and distribution of saponin compounds and their biological activity in various plants. In this chapter, we summarized and discussed the recent progress on saponin biosynthesis pathway and genes involved in the up and downstream pathway of saponins.

2.1 Introduction

Plants are renewable bio-resources, providing a great amount of raw materials especially phytochemicals and lingo cellulosic biomass for various industrial firms, including pharmaceutical, textile and cosmetic sections (Guerriero et al. 2018). However, plants are sessile organisms, and to protect themselves against exogenous constraints, plants usually produce a diverse array of bioactive metabolites with complex chemical compositions in response to different forms of biotic and abiotic stresses (Ncube et al. 2015; Hidalgo et al. 2018; Abdelrahman et al. 2017a, b). Plant bioactive metabolites can be classified into four key major classes, including phenolics, terpenoids, sulphur-containing compounds and alkaloids (Abdelrahman et al. 2018, 2019). These bioactive metabolites exhibited antimicrobial, repellents and/or deterrents properties against wide range of plant pathogenic fungi and bacteria, as well as nematodes and insectivores. Terpenoids are

M. Abdelrahman, S. Jogaiah, *Bioactive Molecules in Plant Defense*,
https://doi.org/10.1007/978-3-030-61149-1_2

diverse and the largest class of naturally occurring plant secondary metabolites including tri, di, sesqui and monoterpenes derived from terpenes, where methyl groups have been removed or moved, or oxygen atoms being added (Davis and Croteau 2000; Ayoola 2008). Terpenoids have numerous functions in basic physiological processes as well as the interaction of plants with their environments. Likewise, steroidal saponins are a group of high molecular weight bioactive compounds present naturally in various plant species. The significance of terpenoids and steroidal saponins is a consequence of their potential pharmacological activity and industrial use as hypo-cholesterolemic, antitumoral, antiplatelet, immune adjuvant, anti-HIVanti-inflammatory, antibacterial, fungicide and anti-leishmanial agents (Yan et al. 2006; Ma et al. 2007; Kuo et al. 2009; Yendo et al. 2010; Mostafa et al. 2013; Abdelrahman et al. 2017a, b, c, d).Terpenoids and steroidal saponins are recognized by a outline derived from 30-carbon (30-C), a oxidosqualene precursor belongs to glycosyl residues and are connected, via the 2-C-methyl-D-erythritol 4-phosphate (MEP) pathway in plastids and/or through the mevalonic acid (MVA) signaling pathway in the cytoplasm (Fig. 2.1) (Rohdich et al. 2001; Haralampidis et al. 2002; Zhao and Li 2018). The cyclization of the precursor 2, 3-oxidosqualene compound through oxidosqualene cyclase (OSC) enzyme, combined with steroidal-skeleton modifications through glycosylation and hydroxylation resultedin the formation of numerous terpenoid and saponin compounds with different compositionsand antifungal properties (Kalra et al. 2013; Mostafa et al. 2013). Despite many studies on the pharmaceutical activities and chemical structure of steroidal saponins and terpenoids, little is known about the downstream level of the molecular mechanisms involved in cyclization process (Abdelrahman et al. 2017a, b, c, d).In this chapter we will discuss the recent literature regarding the production and biosynthesis process of terpenoids and steroidal saponins in wide range of plants, with special focus on steroidal saponin biosynthesis-related genes.

2.2 Biosynthesis of Triterpenoid and Steroidal Saponins

Even though the saponins are considered as the largest natural bio-product members, still, its biological process and applications are not fully understood. Saponins are commonly known for its significant applications in the response mechanisms of plants against pathogens, herbivores and pests, because of their antiparasitic, antimicrobial and insecticidal properties (Moses et al. 2014; Abdelrahman et al. 2017d). Saponins are chemically diverse bioactive compounds comprising either steroidal or triterpenoid aglycones linked with oligosaccharide moieties. Several reports have been documented over the few decades, concentrating on isolation, elucidation of its structural and biological properties of different saponin compounds (Tan et al. 1999; Connolly and Hill 2000; Sparg et al. 2004; Mostafa et al. 2013; Abdelrahman et al. 2017a, b, c, d). These review and research articles provided comprehensive insights into saponin compound structures and classifications. Saponins are frequently grouped into two key classes: (i) steroid saponins and (ii) triterpenoid saponins (Abe et al. 1993; Lanzotti et al. 2012; Abdelrahman et al. 2017c, d), which are

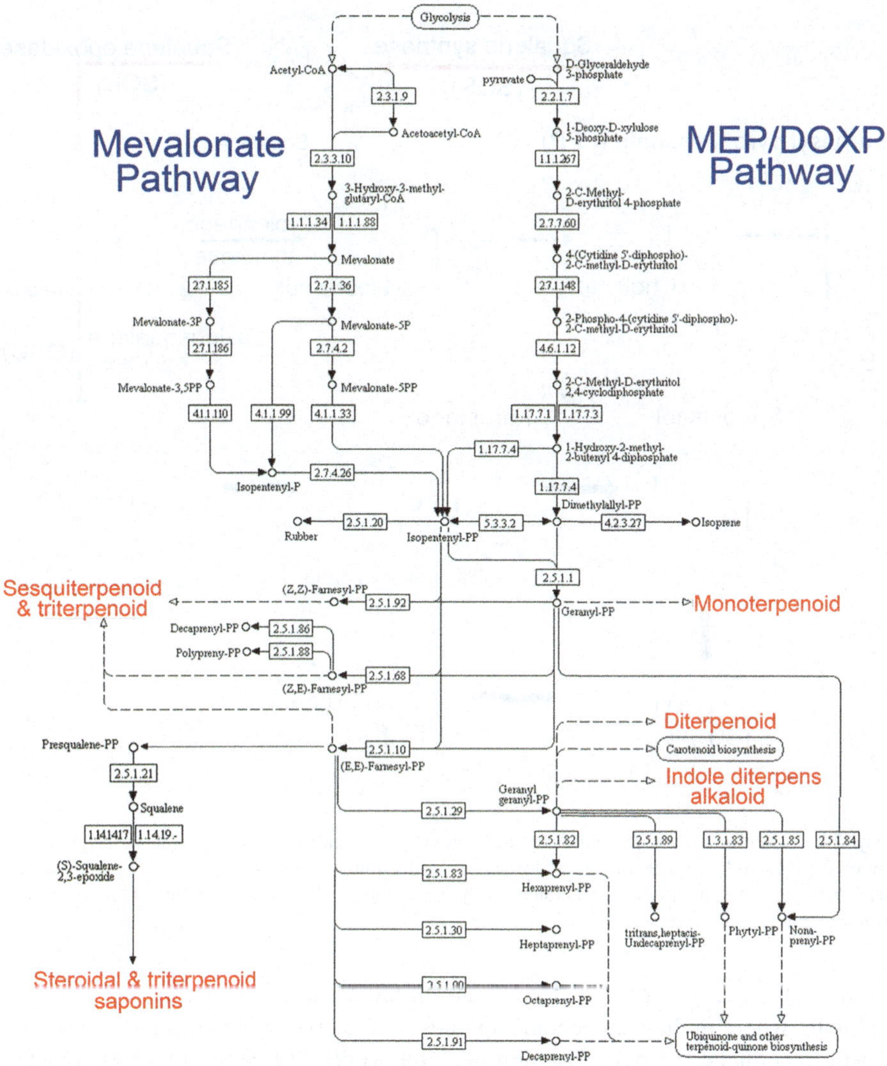

Fig. 2.1 Biosynthesis pathway of terpenoid and steroidal saponins according to Kyoto Encyclopedia of Genes and Genomes (KEGG, www.genome.jp/kegg-bin/show_pathway) database. The saponin and terpenoids are highlighted in red color

developed from the 30-C skeleton-containing oxidosqualene precursor as a key component (Haralampidis et al. 2002; Abdelrahman et al. 2017c). While the biosynthesis of saponin compounds has been reported in various plants, most of the bioactive saponins are produced by the dicotyledonous plant family and they are majorly contains triterpenoid saponin type. Whereas, plant belonging to monocotyledonous species mostly produced the steroidal-type saponins. The main difference

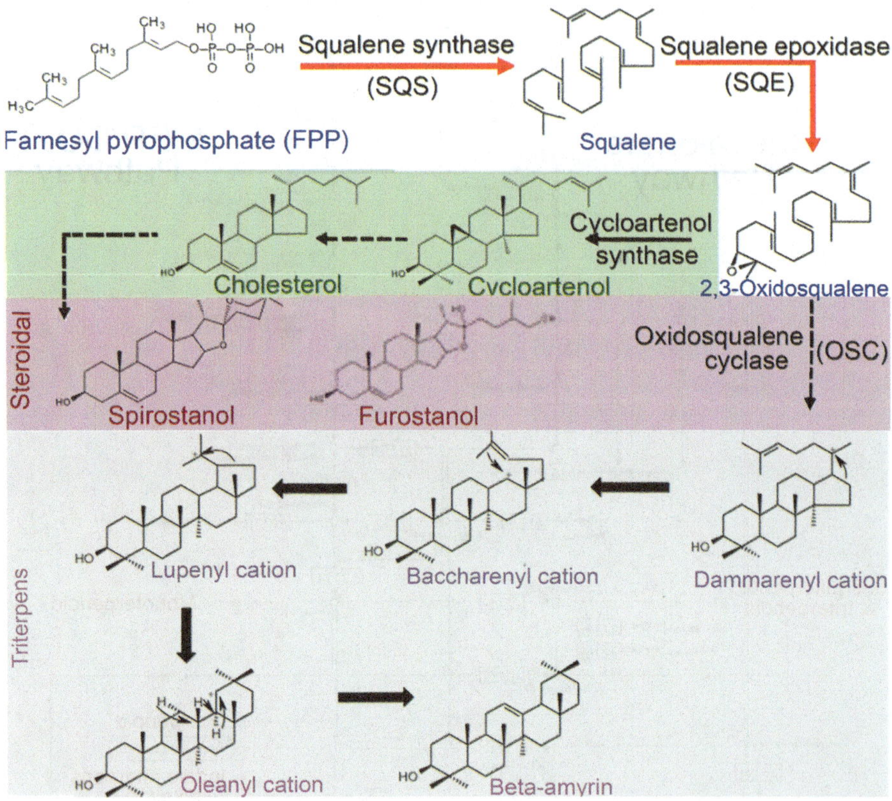

Fig. 2.2 Triterpenoid and steroidal saponins biosynthesis pathway starting from farnesyl diphosphate (FPP). Cholesterol is the devoted predecessor for the synthesis of steroidal saponins (violet). The triterpenoid saponins are generated from 2,3-oxidosqualene that is cyclized to specialized triterpene aglycones (gray)

between the two groups is mainly based on the scenery of the aglycone character strength, where in the triterpenoid saponins all the 30-C atom skeleton is reserved. On the other hand, the steroid saponins exhibited only 27-C atoms by removing three methyl groups (Haralampidis et al. 2002; Sparg et al. 2004; Mostafa et al. 2013; Itkin et al. 2013; Abdelrahman et al. 2017c, d). From a biosynthetic view point, the steroidal and triterpenoid aglycone backbones are generated from isoprenoids through the MVA pathway (Fig. 2.1). In the MVA signaling pathway, the 3-hydroxy-3-methylglutaryl-CoA reductase (HMGR) catalyzes the transformation of acetyl-CoA into the terpene (5-C) precursor isopentyl pyrophosphate (IPP). IPP then isomerized into its allylic isomer dimethyl allyl pyrophosphate (DMAPP) mediated by isopentenyl diphosphate isomerase (IDI) enzyme. The successive concentration of the two IPP moleculeswith one DMAPP molecule through head-to-tail link resulted in the generation of 15-C product farnesyl pyrophosphate (FPP), an intermediate saponin precursor (Fig. 2.2). Subsequently, the concentration of two

FPP moleculesthrough tail-to-tail link mediated by squalene synthase (SQS) produces the linear 30-C squalene precursor. Squalene is further oxidized by the squalene epoxidase (SQE) enzyme to form 2,3-oxidosqualene (Fig. 2.2). The 2,3-oxidosqualene is cyclized by a diversity of oxidosqualene cyclase (OSC)-enzymes into multicyclic structures, which isconsidered as the branching point between metabolisms of steroidal and triterpene saponin. The precursor cycloartenol. a tetracyclic in nature is created via the cyclization of 2,3-oxidosqualene through cycloartenol synthase (CAS) enzyme. In monocot plants, combination of phytosterols are produced from the tetracyclic cycloartenol, including the 29-C sitosterol, the 28-C campesterol and the 27-C cholesterol. Later, the steroidal saponins are synthesized by a sequence of oxido-reduction of aglycone strength before it is glycosylated with other groups of sugar moieties to form furostanol or spirostanol-type saponins with a complex O-heterocycle in their central aglycone structure (Fig. 2.2) (Cammareri et al. 2008; Yendo et al. 2010; Vincken et al. 2007; Abdelrahman et al. 2014, 2017c). Furostanol saponins hold a methyl acetal, hemiacetal, or Δ20 (22)-unsaturation at 22nd C locus, while spirostanol saponins contain a bicyclic spiroacetal moiety at 22^{nd}C that involves the steroid E and F rings (Challinor and De Voss 2013). Similarly, a number of *OSC* genes involved in the development of triterpene C skeletons have been isolated and characterized (Augustin et al. 2011). Also the key triterpene biosynthesis-related genes *β-amyrin synthases* and *lupeol synthases* have been purified and identified in various plants with mono-functional mode, while most of the *α-amyrin synthases* genes isolated so far exhibited broad functional mode and produce more than one triterpene compound (Huang et al. 2012). Unlike OSCs, the identification of novel oxido-reduction cytochrome P450s (CYP450s)- and glycosylation by UDP glycosyltransferases (UGTs)-related genes which are engaged in the formation of triterpenoid group of saponins exhibited, some challenges due to the huge members of CYP450s and UGTs in addition to the weak association between functions and gene similarity of two gene groups (Zheng et al. 2014).

2.3 Candidate Genes Associated with the Biosynthesis Process of Steroidal Saponins

In plants, during the biosynthesis of varied plant secondary metabolites such as fatty acids, terpenoids, hormones, lignins, sterols, and pigments, CP450s catalyze the oxidative reactions while UGTs regulate the glycosylation process, leading to divergence of natural products in plants (Pérez et al. 2013; Abdelrahman et al. 2017c). The diverse features of CP450s and UGTs forms the identity difficult in saponin biosynthesis pathway. For example, The *Arabidopsis thaliana* has a total of 246 CP450 and 112 UGTs species, making investigation of the role of each CP450 and UGT gene by repeal genetics method is a challenge (Werck-Reichhart et al. 2002; Paquette et al. 2003; Nelson 2006). However, such alterations are essential for production the saponin compounds dynamic and soluble in nature. Thus, researchers attempt to isolate many CP450s and UGTs associated in biosynthetic of saponin pathway in different plant species using next-generation sequencing (NGS) approach (Abdelrahman et al. 2017c, d). Plant CP450 can be separated into two major classes, including non-A-type P450 and A-type P450 (Kahn and Durst 2000). The non-A-type P450s are extremely different group that belongs to P450 involved in hormone or lipid metabolism, while the A-type P450s are mostly intricate in secondary metabolites biosynthesis (Paquette et al. 2000). Among these, CYP51 family members are the utmost preserved P450 amongst phyla and exhibit sequence similarity >30–40%. In *Sorghum bicolor*, *Obtusifoliol 14α demethylase* gene which consists to P450 a gene family of CYP51 was isolated and showed stringent substrate specificity towards obtusifoliol (Bak et al. 1997). Wheat *sterol 14α demethylase* gene (CYP51) was also found to harmonize *lanosterol 14-alpha-demethylase* (Cabello-Hurtado et al. 1999). In *Arabidopsis*, *CYP708A2* and CYP705A5 genes were identified as P450 species in triterpenoid thalianol metabolism (Field and Osbourn 2008). In soybean, *CYP93E1 C-24 hydroxylase* was identified as oxidase-related genes in soyasaponin biosynthesis (Shibuya et al. 2006). In *G. glabra*, CYP88D6 was identified as a β-amyrin C-11 oxidase-related gene (Seki et al. 2008). Additionally, CYP93E3 was also identified as a β-amyrin C-24 oxidase-related gene in the secondary metabolism of glycyrrhizin. Han et al. (2011, 2012, 2013) identified three CYP genes, CYP716A47, CYP716A53v2and CYP716A52v2 from *Panax ginseng,* showing that these CYPs are complex in biosynthesis of ginsenoside. Similarly, Zhao et al. (2019) identified 100 PgCYP genes, whose expressions study was highly associated with the contents of ginsenoside. In a recent study by Abdelrahman et al. (2017c) using *Allium fistulosum-A. cepa* Aggregatum group, monosomic addition line demonstrated that *CYP734A1*, *CYP72B1*, *CYP71B31* and *CYP94C1* were strongly upregulated in *A. fistulosum* with addition chromosome 2A from shallot which was attributed to the biosynthesis of steroidal saponin Alliospiroside A.

Glycosylation of saponin compounds is invented as the final step that forms the synthesis of saponins. Glycosylation is too important factor for increasing water solubility and biological activities of saponins (Sawai and Saito 2011; Lanzotti et al. 2012). Therefore, isolation andcharacterization of these UGT enzymes that catalyzed the transport of sugar moieties to these terpenoid and steroidal saponins will positively enable us to better understand the diversity and mechanism of the biological activities of these saponins in different plants. However, only few UGTs connected in biosynthesis of saponin was well identified. Abdelrahman et al. (2017c) reported that *UGT73B5*, *UGT71B1*, and *UGT73C6* expression was attributed to Alliospiroside A saponin biosynthesis in *A. fistulosum-A. cepa* aggregatum group. Ma et al. (2016) reported that several unigenes (comp20876 c0 and comp18634 c0) and (comp18634 c0 and comp20876 c0) isolated from transcriptome analysis of *P. grandiflorum* were greatly homologous, respectively to *Barbarea vulgaris* UGT73C10 and UGT73C11, which converts sapogenin 3-O-glucosylation (Augustin et al. 2012) and *Saponaria vaccaria* UGT74M1, that is a glucosyltransferase belongs to triterpene carboxylic acid (Meesapyodsuk et al. 2007), indicating that unigenes have the similar role in the formation of triterpenoid group of saponins in *Platycodon grandiflorum.* In ginseng, only few UGT members have been categorized for their saponin substrate specificity, including PgUGT74AE2, PgUGT71A27, PgUGT94Q2, UGTPg101 and UGTPg100 (Jung et al. 2014; Yan et al. 2014; Wei et al. 2015), whereas many other UGT members associated with ginsenoside biosynthesis still imperative. Considering the fast progress in NGS technologies, hopefully more members of UGTs would be identified and characterized for their role in saponin biosynthesis.

References

Abdelrahman M, Hirata S, Ito S-I, Yamauchi N, Shigyo M (2014) Compartmentation and localization of bioactive metabolites in different organs of *Allium roylei*. Biosci Biotechnol Biochem 78:1112–1122

Abdelrahman M, El-Sayed M, Jogaiah S, Burritt DJ, Tran LP (2017a) The "STAY-GREEN" trait and phytohormone signaling networks in plants under heat stress. Plant Cell Rep 36:1009–1025

Abdelrahman M, Burritt DJ, Tran LP (2017b) The use of metabolomic quantitative trait locus mapping and osmotic adjustment traits for the improvement of crop yields under environmental stresses. Semin Cell Dev Biol 83:86–94

Abdelrahman M, El-Sayed M, Sato S, Hirakawa H, Ito S-I, Tanaka K, Mine Y, Sugiyama N, Suzuki Y, Yamauchi N, Shigyo M (2017c) RNA-sequencing-based transcriptome and biochemical analyses of steroidal saponin pathway in a complete set of *Allium fistulosum-A. cepa* monosomic addition lines. PLoS One 12:e0181784

Abdelrahman M, Mahmoud HYAH, El-Sayed M, Tanaka S, Tran LS (2017d) Isolation and characterization of Cepa2, a natural alliospiroside A, from shallot (*Allium cepa* L. Aggregatum group) with anticancer activity. Plant Physiol Biochem 116:167–173

Abdelrahman M, El-Sayed MA, Hashem A, Abd_Allah EF, Alqarawi AA, Burritt DJ, Tran LP (2018) Metabolomics and transcriptomics in legumes under phosphate deficiency in relation to nitrogen fixation by root nodules. Front Plant Sci 9:922

Abdelrahman M, Hirata S, Sawada Y, Hirai MY, Sato S, Hirakawa H, Mine Y, Tanaka K, Shigyo M (2019) Widely targeted metabolome and transcriptome landscapes of *Allium fistulosum–A. cepa* chromosome addition lines revealed a flavonoid hot spot on chromosome 5A. Sci Rep 9:3541

Abe I, Rohmer M, Prestwich GC (1993) Enzymatic cyclization of squalene and oxidosqualene to sterols and triterpenes. Chem Rev 93:2189–2206

Augustin JM, Kuzina V, Andersen SB, Bak S (2011) Molecular activities, biosynthesis and evolution of triterpenoidsaponins. Phytochemistry 72:435–457

Augustin JM, Drok S, Shinoda T, Sanmiya K, Nielsen JK, Khakimov B et al (2012) UDP-glycosyltransferases from the UGT73C subfamily in *Barbarea vulgaris* catalyze sapogenin 3-O-glucosylation in saponin-mediated insect resistance. Plant Physiol 160:1881–1895

Ayoola GA (2008) Phytochemical screening and antioxidant activities of some selected medicinal plants used for malaria therapy in southwestern Nigeria. Trop J Pharm Res 7:1019–1024

Bak S, Kahn RA, Olsen CE, Halkier BA (1997) Cloning and expression in *Escherichia coli* of the *obtusifoliol 14α-demethylase* of *Sorghum bicolor* (L.) Moench, a cytochrome P450 orthologous to the *sterol 14α-demethylases* (*CYP51*) from fungi and mammals. Plant J 11:191–201

Cabello-Hurtado F, Taton M, Forthoffer N, Kahn R, Bak S, Rahier A, Werck-Reichhart D (1999) Optimized expression and catalytic properties of a wheat *obtusifoliol 14α-demethylase* (*CYP51*) expressed in yeast. Complementation of erg11Delta yeast mutants by plant CYP51. Eur J Biochem 262:435–446

Cammareri M, Consiglio MF, Pecchia P, Corea G, Lanzotti V, Iebas JI, Tavae A, Conicella C (2008) Molecular characterization of β-amyrin synthase from *Aster sedifolius* L. and triterpenoid saponin analysis. Plant Sci 175:255–261

Challinor VL, De Voss JJ (2013) Open-chain steroidal glycosides, a diverse class of plant saponins. Nat Prod Rep 30:429–454

Connolly JD, Hill RA (2000) Triterpenoids. Nat Prod Rep 17:463–482

Davis EM, Croteau R (2000) Cyclization enzymes in the biosynthesis of monoterpenes, sesquiterpenes, and diterpenes. Top Curr Chem 209:53–95

Field B, Osbourn AE (2008) Metabolic diversification-independent assembly of operon-like gene clusters in different plants. Science 320:543–547

Guerriero G, Berni R, Muñoz-Sanchez JA, Apone F, Abdel-Salam EM, Qahtan AA, Alatar AA, Cantini C, Cai G, Hausman J-F, Siddiqui KS, Hernández-Sotomayor SMT, Faisal M (2018) Production of plant secondary metabolites: examples, tips and suggestions for biotechnologists. Genes 9:309

Han JY, Kim HJ, Kwon YS, Choi YE (2011) The Cyt P450 enzyme CYP716A47 catalyzes the formation of protopanaxadiol from dammarenediol-II during ginsenoside biosynthesis in *Panax ginseng*. Plant Cell Physiol 52:2062–2073

Han JY, Hwang HS, Choi SW, Kim HJ, Choi YE (2012) Cytochrome P450 CYP716A53v2 catalyzes the formation of protopanaxatriol from protopanaxadiol during ginsenoside biosynthesis in *Panax ginseng*. Plant Cell Physiol 53:1535–1545

Han JY, Kim MJ, Ban YW, Hwang HS, Choi YE (2013) The involvement of β-amyrin 28-oxidase (CYP716A52v2) in oleanane-type ginsenoside biosynthesis in *Panax ginseng*. Plant Cell Physiol 54:2034–2046

Haralampidis K, Trojanowska M, Osbourn AE (2002) Biosynthesis of triterpenoid saponins in plants. Adv Biochem Eng Biotech 75:31–49

Hidalgo D, Sanchez R, Lalaleo L, Bonfill M, Corchete P, Palazon J (2018) Biotechnological production of pharmaceuticals and biopharmaceuticals in plant cell and organ cultures. Curr Med Chem 25:3577–3596

Huang L, Li J, Ye H, Li C, Wang H, Liu B, Zhang Y (2012) Molecular characterization of the pentacyclictriterpenoid biosynthetic pathway in *Catharanthusroseus*. Planta 236:1571–1581

Itkin M, Heinig U, Tzfadia O, Bhide AJ, Shinde B, Cardenas PD, Bocobza SE, Unger T, Malitsky S, Finkers R, Tikunov Y, Bovy A, Chikate Y, Singh P, Rogachev I, Beekwilder J, Giri AP, Aharoni A (2013) Biosynthesis of antinutritional alkaloids in solanaceous crops is mediated by clustered genes. Science 341:175–179

Jung SC et al (2014) Two ginseng UDP-Glycosyltransferases synthesize Ginsenoside Rg(3) and Rd. Plant Cell Physiol 55:2177–2188

Kahn RA, Durst F (2000) Function and evolution of plant cytochrome P450. Recent Adv Phytochem 34:151–189

Kalra S, Puniya PL, Kulshreshtha D, Kumar S, Kaur J, Ramachandran S, Singh K (2013) De novo transcriptome sequencing reveals important molecular networks and metabolic pathways of the plant, *Chlorophytum borivilianum*. PLoS One 8:e83336

Kuo RY, Qian K, Morris-Natschke SL, Lee KH (2009) Plant-derived triterpenoids and analogues as antitumor and anti-HIV agents. Nat Prod Rep 26:1321–1344

Lanzotti V, Romano A, Lanzuise S, Bonanomi G, Scala F (2012) Antifungal saponins from bulbs of white onion, *Allium cepa* L. Phytochemistry 74:133–139

Ma YX, Fu HZ, Li M, Sun W, Xu B, Cui JR (2007) An anticancer effect of a new saponin component from *Gymnocladus chinensis* Baillon through inactivation of nuclear factor-kappa B. Anti-Cancer Drugs 18:41–46

Ma C-H, Gao Z-J, Zhang J-J, Zhang W, Shao J-H, Hai M-R, Chen J-W, Yang S-C, Zhang G-H (2016) Candidate genes involved in the biosynthesis of triterpenoid saponins in *Platycodon grandiflorum* identified by transcriptome analysis. Front Plant Sci 7:673

Meesapyodsuk D, Balsevich J, Reed DW, Covello PS (2007) Saponin biosynthesis in *Saponaria vaccaria*. cDNAs encoding *β-amyrin synthase* and a triterpene carboxylic acid *glucosyltransferase*. Plant Physiol 143:959–969

Moses T, Papadopoulou KK, Osbourn A (2014) Metabolic and functional diversity of saponins, biosynthetic intermediates and semi-synthetic derivatives. Crit Rev Biochem Mol Biol 49:439–462

Mostafa A, Sudisha J, El-Sayed M, Ito S-I, Ikeda T, Yamauchi N, Shigyo M (2013) Aginosidesaponin, a potent antifungal compound, and secondary metabolite analyses from *Allium nigrum* L. Phytochem Lett 6:274–280

Ncube B, Staden V, Tilting J (2015) Plant metabolism for improved metabolite biosynthesis and enhanced human benefit. Molecules 20:12698–12731

Nelson D (2006) Plant cytochrome P450s from moss to poplar. Phytochem Rev 5:193–204

Paquette SM, Bak S, Feyereisen R (2000) Intron–exon organization and phylogeny in a large superfamily, the paralogous cytochrome P450 genes of *Arabidopsis thaliana*. DNA Cell Biol 19:307–317

Paquette S, Møller BL, Bak S (2003) On the origin of family 1 plant glycosyltransferases. Phytochemistry 62:399–413

Pérez AJ, Calle JM, Simonet AM, Guerra JO, Stochmal A, Macías FA (2013) Bioactive steroidal saponins from *Agave offoyana* flowers. Phytochemistry 95:298–307

Rohdich F, Kis K, Bacher A, Eisenreich W (2001) The nonmevalonate pathway of isoprenoids: genes, enzymes and intermediates. Curr Opin Chem Biol 5:535–540

Sawai S, Saito K (2011) Triterpenoid biosynthesis and engineering in plants. Front Plant Sci 2:25

Seki H, Ohyama K, Sawai S, Mizutani M, Ohnishi T, Sudo H, Akashi T, Aoki T, Saito K, Muranaka T (2008) Licorice *β-amyrin 11-oxidase*, a cytochrome P450 with a key role in the biosynthesis of the triterpene sweetener glycyrrhizin. Proc Natl Acad Sci USA 105:14204–14209

Shibuya M, Hoshino M, Katsube Y, Hayashi H, Kushiro T, Ebizuka Y (2006) Identification of *β-amyrin* and *sophoradiol 24-hydroxylase* by expressed sequence tag mining and functional expression assay. FEBS J 273:948–959

Sparg SG, Light ME, Staden JV (2004) Biological activities and distribution of plant saponins. J Ethnopharmacol 94:219–243

Tan N, Zhou J, Zhao S (1999) Advances in structural elucidation of glucuronideoleanane-type triterpene carboxylic acid 3,28-O-bisdesmosides (1962–1997). Phytochemistry 52:153–192

Vincken JP, Heng L, Groot A, Gruppen H (2007) Saponins, classification and occurrence in the plant kingdom. Phytochemistry 68:275–297

Wei W, Wang P, Wei Y, Liu Q, Yang C, Zhao G, Yue J, Yan X, Zhou Z (2015) Characterization of *Panax ginseng* UDP-Glycosyltransferases catalyzing protopanaxatriol and biosyntheses of bioactive ginsenosides F1 and Rh1 in metabolically engineered yeasts. Mol Plant 8:1412–1424

Werck-Reichhart D, Bak S, Paquette S (2002) Cytochrome P450. Arabidopsis Book 1:e0028

Yan MC, Liu Y, Chen H, Ke Y, Xu QC, Cheng MS (2006) Synthesis and antitumor activity of two natural N-acetylglucosamine-bearing triterpenoidsaponins: lotoidoside D and E. Bioorg Med Chem Lett 16:1200–1204

Yan X, Fan Y, Wei W, Wang P, Liu Q, Wei Y, Zhang L, Zhao G, Yue J, Zhou Z (2014) Production of bioactive ginsenoside compound K in metabolically engineered yeast. Cell Res 24:770–773

Yendo ACA, de Costa F, Gosmann G, Fett-Neto AG (2010) Production of plant bioactive triterpenoid saponins: elicitation strategies and target genes to improve yields. Mol Biotechnol 46:94–104

Zhao YJ, Li C (2018) Biosynthesis of plant triterpenoid saponins in microbial cell factories. J Agric Food Chem 66:12155–12165

Zhao M, Lin Y, Wang Y, Li X, Han Y, Wang K, Sun C, Wang W, Zhang M (2019) Transcriptome analysis identifies strong candidate genes for ginsenoside biosynthesis and reveals its underlying molecular mechanism in *Panax ginseng* C.A. Meyer. Sci Rep 9:615

Zheng X, Xu H, Ma X, Zhan R, Chen W (2014) Triterpenoid Saponin biosynthetic pathway profiling and candidate gene mining of the *Ilex asprella* root using RNA-Seq. Int J Mol Sci 15:5970–5987

References

Metabolic and Functional Diversity of Saponins

3

Abstract

The 'saponin' word is originated from Latin name 'sāpō' means 'soap', as saponins make foams when they are shaken using water. These are a varied class of surface active and nonvolatile secondary metabolites are broadly dispersed in nature, existing in diverse species of plants, including both monocot and dicot. Saponins are 30-carbon skeleton molecules derived from oxidosqualene precursor that consisted of nonpolar aglycones, to which one or more polar monosugar molecules are attached. The polar (sugar moieties) and nonpolar (aglycones) structures mixture in the saponin compounds describe their soap like behavior in water and provide the base for their biological activities. Although saponin is considered major group of plant natural products, their functions in plant biological process are not fully understood and saponins are usually recognized to have significant functions in plant defense mechanisms against pathogens, herbivores and pests. Saponin compounds have a wide array of characters, such as emulsifying and foaming, bitterness and sweetness, antimicrobial, insecticidal, as well as pharmacological and medicinal properties. Although in the early times it may be suitable to categorize saponin compounds according to their biological and/or physicochemical activities, currently with the high throughput in chemistry and mass spectrometry, the structural diversity of saponin compounds became the main classification scheme. In this chapter, we will try to describe the different types of saponin compounds and their distributions in the different plant species. The new isolated saponin compounds from different plants will also be listed as a source information for future biological studies.

M. Abdelrahman, S. Jogaiah, *Bioactive Molecules in Plant Defense*,
https://doi.org/10.1007/978-3-030-61149-1_3

3.1 Classification of Saponins

From biological point of view, plant saponins are known as broad spectrum protective compounds against different types of herbivores and pathogenic fungi (Sparg et al. 2004; Szakiel et al. 2011; Abdelrahman et al. 2017a, b). Saponin molecules are usually divided to two key classes, the triterpene saponins and the steroid saponins (Abe et al. 1993; Kuzina et al. 2009; Mostafa et al. 2013a, b; Abdelrahman et al. 2014, 2017a), and both of these two classes are generated from the 30-carbon (C) predecessor oxidosqualene (Abe et al. 1993; Haralampidis et al. 2002; Mostafa et al. 2013a, b). The major variance between the steroid saponins and the triterpene saponins is that triterpene saponins possessed all 30 C-atoms in their skeleton, while in the steroidal saponins, three methyl groups removed from the skeleton and thus their chemical structure consisted of 27 C-atoms (Fig. 3.1). However, Sparg et al. (2004) have also categorized saponins into three categories, such as, the spirostanol, triterpenoid, and furostanol saponins. This classification highlights secondary structures due to secondary bio-transformations, but not due to the biosynthetic point of view. Additionally, glycosteroid alkaloids are also considered as a branch of saponins because these glycosteroid alkaloids exhibit a similar biosynthetic progenitor like saponins, and also comprised monosaccharide moieties attached to a steroidal-type backbone (Haralampidis et al. 2002; Patel and Savjani 2015). However, glycosteroid alkaloids encompass nitrogen (N) atom as an essential, typical portion of their aglycone structure, these separates them into a different classes. From structure point of view, hydrophilic sugar moieties and the hydrophobic aglycone backbone together makes the saponin molecules highly amphipathic and deliberates emulsifying and foaming properties and provide the base for a varied varieties of bioactivities, comprising antimicrobial and insecticidal as well as pharmacological properties (Moses et al. 2014).

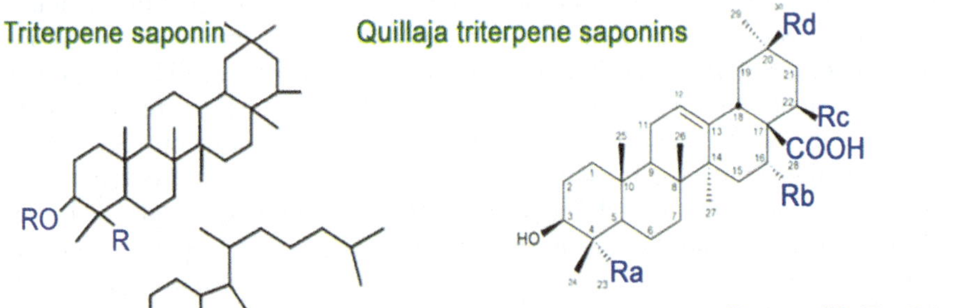

		Ra	Rb	Rc	Rd
Quillaic acid	Q	CHO	OH	H	CH_3
Quillaic acid, 22β-OH	Q-OH	CHO	OH	OH	CH_3
Gypsogenin	G	CHO	H	H	CH_3
Phytolaccinic acid	P	CH_2OH	H	H	$COOCH_3$
Phytolaccinic acid, 23-O-Ac	P-Ac	CH_2OCOCH_3	H	H	$COOCH_3$
Echynocystic acid	E	CH_3	OH	H	CH_3

Fig. 3.1 Schematic view of the steroidal and triterpene saponins structure and *Quillaja* triterpene saponins

Saponins found in domesticated and wild plants in both, with the triterpenoid type saponins are mainly found in *Amaranthaceae, Apiaceae, Araliaceae, Aquifoliaceae, Myrsinaceae, Berberidaceae, Caryophyllaceae, Cucurbitaceae, Zygophyllaceae, Chenopodiaceae,* and *Leguminosae,* families (Shi et al. 2004; Sparg et al. 2004; Parente and Da Silva 2009). On the other hand, the steroidal-type saponins being predominantly present in *Amaryllidaceae*, *Alliaceae*, *Asparagaceae*, *Agavaceae*, *Bromeliaceae*, *Liliaceae*, *Dioscoreaceae*, *Scrophulariaceae* and *Palmae* families (Osbourn 2003; Augustin et al. 2011; Moses et al. 2014). Many plants secrete saponin compounds from rhizosphere soil, these saponins act as allelopathic compounds that inhibit the development of other adjacent plants and also serve as antifungal chemical barrier against soil borne pathogens. For instance, the yield and growth of cotton (*Gossypium barbadense*) and wheat (*Triticum aestivum*) was showed lower when they cultivated on a soil formerly used for *Medicago sativa* farming, compared with growth on uncultivated plain soil (Leshem and Levin 1978; Oleszek 1993). Meta-transcriptomics based study of wheat, pea, and oat plants rhizosphere and bulk soil demonstrated big differences between their microbiomes, and pea rhizosphere was augmented with fungi, compared with the oat rhizosphere, which possessed avenacins as potent antifungal compound. Similarly, the comparison of wild-type (WT) plants and the microorganisms from oat rhizosphere as *sad1* mutants (deficient of avenacins) displayed large difference between fungal and nematode populations, indicating a broader role for avenacins as signaling molecules instead of defensive from pathogenic fungi (Turner et al. 2013).

3.1.1 Quillaja Triterpene Saponins

Quillaja saponaria extracts represent the key source of triterpene saponin for different industrial applications, specifically as immune-stimulant and vaccine-adjuvant, which has resulted in crucial investigation in the field of vaccine progress (Fleck et al. 2019). Saponin compound(s), alone or combined into immune-stimulating multiplexes were able to regulate immunity system by triggering lymphocyte production through cytotoxic T (Th1) and cytokines (Th2), and increasing antigen uptake in response to various antigens (Cibulski et al. 2016; Fleck et al. 2019). In addition, saponins isolated from extracts of *Quillaja saponaria* are widely applied as emulsifiers in processes containing flavors or lipophilic colors, for removal of dietary cholesterol, and as foaming agents in cosmetics preservatives and carbonated beverages (San Martín and Briones 1999; Güçlü-Ustündağ and Mazza 2007; Moses et al. 2014). For examples, a combination of *Quillaja* saponin extracts and lecithins or proteins can enhance the additive emulsifying or synergistic effects and thus preserve the outstanding emulsifying characteristics in food process (Reichert et al. 2019). The most described aglycone from *Q. saponaria*, is quilaic acid (Fig. 3.1), which contains aldehyde functional group from the carbon position C-23, in addition to side chain of acyl as typical features of saponins with *Quillaja*-derived, and these characters play an important roles in their biological activities (Higuchi et al. 1988; Nord and Kenne 2000; Sun et al. 2009). Usually, *Quillaja* bark saponin compounds have been applied as a cleaner and more recently it has been approved for use as food additives, including Japan and China, European Union and the United States (Wojciechowski 2013) and World Health Organization (www.who.int/foodsafety/). Lately, food components possessing *Quillaja* bark saponins (Q-Naturale) were also permitted via the Food and Drugs Administration (FDA) as an effective emulsifier for beverages (de Faria et al. 2017). Besides *Quillaja* saponin applications as emulsifiers and detergents, other pharmacological properties include antifungal, antiviral, antiparasitic and antibacterial have been widely reported (Pen et al. 2006; Roner et al. 2007, 2010; Holtshausen et al. 2009; Dixit et al. 2010; Tam and Roner 2011). *Q. saponaria* raw material has been listed in the Brazilian and European Pharmacopoeias as anti-inflammatory, cough reliever, expectorant, hemolytic and hypo-cholesterolemic agent. Triterpenes are similarly available in the leaf tissues of *Q. brasiliensis* that denotes an alternative renewable source of saponins rather than the destructive and non-renewable use of the barks of *Q. saponaria* tree. Thus, the traditional use of native forests as source of saponins has been recently substituted, and other alternate sources like synthetic analogues of QS-21 saponin became available (Fleck et al. 2006; de Costa et al. 2016; Cibulski et al. 2016). However, search for new natural saponins remains a hot research topic. The first chemical structure elucidation of *Quillaja* triterepen saponins was reported by Higuchi et al. (1988), who identified two deacetylated saponin compounds from *Q. saponaria* bark (Fig. 3.1). Later on isolation and identification of ~60 saponin compounds from *Q. saponaria*, that contained β-amyrin-derived triterpen sapogning/aglycone, especially quillaic acid, have been also reported (Guo et al. 1998; Guo and Kenne 2000; Nord and Kenne 2000; Nyberg et al. 2003). These *Quillaja* triterpene saponins are glycosylated by disaccharide residues at two positions C-28 and C-3 sites in the sapogenin skeleton. The aglycone from C-3 site is commonly glycosylated through β-d- Galp-(1 → 2)-β-d-GlcAp, which usually divided at position O-3 of *β*-d-Xylpor *α*-l-Rhap residue with glucuronic acid. In limited cases, some saponins lack the C-28 linked oligosaccharide, but in most common scenario C-28 location is glycosylated via acetal-ester linkage to an oligosaccharide. The generated oligosaccharide has a preserved residue consisted of theα-l-Rhap-(1 → 2) -β-d- Fucp disaccharide in various tailors (Kite et al. 2004; Fleck et al. 2019). In addition, the *β*-d-Fucp glucoside moiety attached at C-28 position is acylated from O-4 position; however, the equivalent region isomers acylated from O-3 site can be detected in solution because of *trans*-esterification reactions occurred between the hydroxyl groups of *cis*-vicinal, which were detected in the extra saponin QS-21 compound (Jacobsen et al. 1996; Fleck et al. 2019). In addition to the saponin, phenolic compounds in aqueous, *Quillaja* saponin extract indicated that piscidic acid and *p*-coumaric acid derivatives represent the major constituents, while other phenolic contents such as vanillic acid and glucosyringic acid derivatives were identified at lower level (Maier et al. 2015). The phenolic compounds in the *Quillaja* saponins might contribute to the antioxidant activity as well as undesired browning reactions in the final product in *Q. saponaria* extracts.

3.1.2 Ginseng Triterpene Saponins

Ginseng (*Panax ginseng*) is an enduring medicinal plant; it belongs to the *Araliaceae* family is one of the main active ingredients in the traditional Asian herbal remedies (Zhu et al. 2004; Shin et al. 2015; Lee et al. 2017). 'Ginseng' name is derived from a Chinese word "man-like" shape of the root. With a total production of ~80,000 tons which worth nearly $2.1 billion market share in 2013, ginseng plants and its by-products occupy a conspicuous rank of bestselling natural produces in the worldwide, and worldwide ginseng marketplace is anticipated to be value $7.51 billion by 2025 (Baeg and So 2013). Thus, fast growth of technology has advanced several features of ginseng study (Qi et al. 2011). The two utmost familiar plant species are *P. ginseng* (usually recognized as Korean/Chinese ginseng) and *P. quinquefolius* (American ginseng). While, other low recognized plant species in genus ginseng, including Vietnamese ginseng *P. vietnamensis* Ha et Grushv and Japanese ginseng *P. japonicus* can be found. The ginseng triterpene saponins were initially purified by using thin layer chromatography (TLC) and named alphabetically based on their position on the TLC system as follow: Rb1, Rb2, Re, Rc, and etc. (Kaku et al. 1975). The main compounds of ginseng saponins contains a 4ring scheme by *trans* association, and various sugar residues, including glucose, xylose, rhamnose and arabinose connected to the C3, C6 and C20 positions (Shin and Oh 2016; Lu et al. 2019). Then, every ginseng saponin compound can be additionally categorized based on Dammarane structure, comprising the panaxadiol type (PPD), that compromises a H atom from C6 (Rc, Rd., Rs, Rb1, Rg3, Rb2, Rb3, Rh2,); panaxatriol type (PPT), which compromises a C6 sugar side chain (Re, Rh, Rg1, Rf, Rg2); or oleanane type, these exhibits 2 minor compounds namely Ro and F11 (Wong et al. 2015; Shi et al. 2019). Ginseng saponins, such as Rc, Rb1, Re, Rb2, Rd., Rf, and Rg1are the maximum abundant in ginseng roots, compromising more than 90% of the total saponin contents in *P. ginseng* species (Mohanan et al. 2018; Shi et al. 2019). Rf and Rg1 are extremely concentrated in the barks and interior central parts of the ginseng roots, while Re, Rd. and Ra1, are more distributed and found in bark but very low in the inner core of the root (Shi et al. 2019). Also Rb1 can be found in the roots, root hairs and rhizomes of ginseng, relative to leaves and stems (Xu et al. 2017). The key three OSC enzymes such as, cycloartenol synthase (CAS), β-amyrin synthase (BAS), dammarenediol-II synthase (DS), and catalyze the conversion of 2,3-oxidosqualene into 3 different intermediate constituents namely β-amyrin, dammarenediol-II, and cycloartenol (Fig. 3.2). The β-amyrin and Dammarenediol-II are eventually converted to ginseng saponins (Tansakul et al. 2006; Shin et al. 2015). However, dammarenediol-II is the main component of dammarane type ginseng saponins, containing ginsenosides Rb2, Rb1, Rg1, and Re, which explanation for higher percentage of saponin compounds identified in species of ginseng. In contrast, oleanane type triterpene saponins synthesized from β-amyrine, are very infrequent and frequently unnoticeable in *P. ginseng* saponins, except ginsenoside Ro (Qi et al. 2011; Kwon 2019). Dammarane-type saponins are additional categorized into several groups, including PPT and PPD saponins (Hong et al. 2009; Yang et al. 2018; Kochan et al. 2019). The more than 20 enzymatic steps

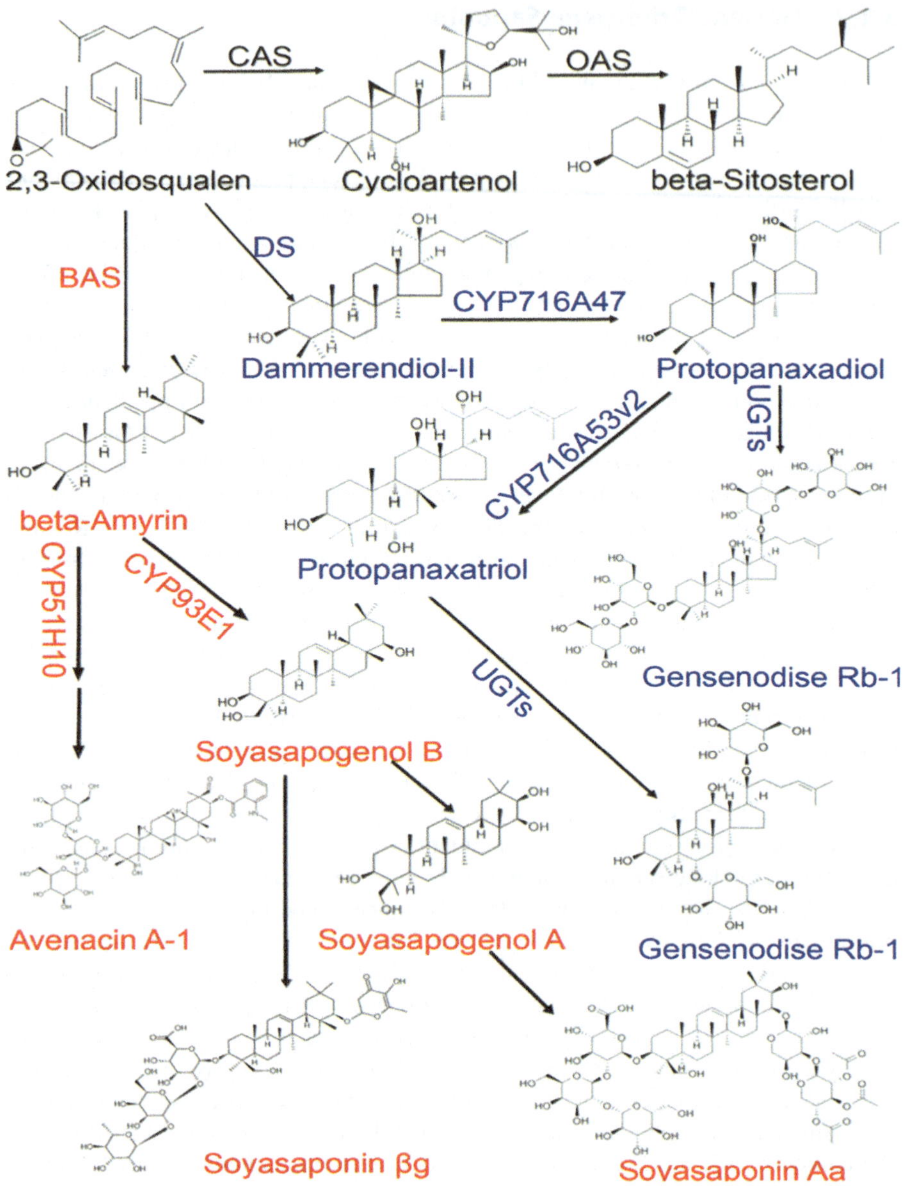

Fig. 3.2 A schematic view of ginseng triterpene saponin pathway. Cycoartenol synthase, CAS; *β*-amyrine synthase, BAS; dammarenedino-II synthase, DS

in ginsenosides biosynthetic pathway, including a sequence of crucial enzymes like squalene synthase (SS), farnesyl pyrophosphate synthase (FPS), 3- hydroxy-3-methylglutaryl coenzyme A reductase (HMGR), DS, CYP450, squalene epoxidase (SQE), BAS, and UDP-glycosyltransferase (UGT) are mainly intricate

(Kim et al. 2014; Yang et al. 2018). For instance, OSCs catalyze 2, 3-oxidosqualene cyclization, which can produce above 100 types of triterpenoids with various frames.

Many of *OSC* genes derived from crops, medicinal and model plants have been cloned and functionally characterized. For instance, *Arabidopsis* genome contained 13 *OSC* genes and all of them have been functionally characterized (Kushiro and Ebizuka 2010). Two *OSC* genes namely *BAS* and *DS* were produced from *P. ginseng* and known to be intricate in production of oleanane-type ginsenosides, and dammarane-type respectively (Kushiro et al. 1997; Tansakul et al. 2006). The in vitro assay by Kushiro et al. (1997) using furry roots of *P. ginseng* demonstrated that DS catalyze the conversion of dammarendiol-II from 2,3- oxidosqualene. Later on, Tansakul et al. (2006) reported that *DS* is the key gene for ginseng saponin using lanosterol synthase deficient (*erg7*) yeast strain GIL77. Similarly, Han et al. (2006) successfully cloned a *DS* gene termed DDS isolated from *P. ginseng* flower into yeast transformant and they were able to detect dammarendiol-II and hydroxyl dammarenone as products, and the exogenous application of Methyl jasmonate (MeJA) can induced the *DDS* gene expression level from *P. ginseng* roots. Instead, stopping of *DDS* gene using RNAi in transgenic *P. ginseng* reduced the ginsenoside production percentage by 84.5% in the roots, however, the silencing of *DDS* gene also induced the expression of other *OSCs* genes, suggesting that close crosstalk between *DDS* gene with other OSCs because both utilize the similar precursor (Lee et al. 2011). Above reports suggested that the induction of *DDS* gene expression displays a dynamic function in *P. ginseng* for the biosynthesis process of ginsenosides, and therefore dammarenediol-II generation is recognized as a rate limiting stage. Therefore, over expression of *DDS* gene from *P. ginseng* can raise the secretion of ginsenosides, which would increase the quality of *P. ginseng*, because ginseng harbor great dammarane-type ginsenoside is considered high mark (Luo et al. 2011; Niu et al. 2014). After an OSC builds the simple triterpenoid backbone, it is being altered into a hydrophobic aglycone named sapogenin. The initial step in saponin modification is oxidation through P450, which is followed by further modification reaction such as O-glycosylation by UGT (Kahn and Durst 2000).

Glycosylation process has been well known as a vital step for saponin biosynthesis and increasing its solubility in water, and subsequently affects the related biological activity of triterpene saponins. Both P450 and UGT species are belonging to big gene families, and both of them are the main aspects for divergence of several natural products, including saponins in different plant species, which make the identification process of saponin-specific P450 and UGT genes difficult (Sawai and Saito 2011). The conventional ginseng saponin extraction method uses Soxhlet, heat-reflux, shaking and/or ultrasound-assisted extraction (UAE) (Qi et al. 2011; Jegal et al. 2019). Recently, newer automated methods using less solvent and take short time with high efficiency such as pressurized hot water extraction (PHWE), microwave assisted extraction (MAE), high-pressure MAE, pressurized liquid extraction (PLE), and supercritical fluid extraction (SFE) had been recently used for ginseng saponin extractions (Qi et al. 2011; Jegal et al. 2019). List of the ginseng saponin compounds that have been isolated from different organs in different Panax species are summarized in Table 3.1.

3.1.3 Soybean Triterpene Saponins

Soybeans have long been recognized as an excellent source not only for high-quality proteins but also for a wide range of bioactive compounds such as saponins, isoflavones, oils and fibers (MacDonald et al. 2005; Sugiyama 2019). Many kinds of soyasaponins are present in soybean seed, which can be classified based on their chemical structure into 4 main classes according to the aglycone, including, glycosides of soyasapogenol E, soyasapogenol B, soyasapogenol A and soyasapogenol B, the C22 bound to 2,3-dihydro-2,5-dihydroxy-6-methyl-4H-pyran-4-one (DDMP) remains (Shiraiwa et al. 1991a; Kitagawa et al. 1998; Berhow et al. 2000, 2002; Sugiyama 2019). It is recognized that the 'B-type saponins' are very wide in seeds of edible soybeans, while 'A-type saponins' are found mainly in soybean seed hypocotyl. However, the 'E-type saponins' are mostly unstable and serve as a precursor of 'B-type saponins' (Shiraiwa et al. 1991a, b). The composition of soyasaponins diverse with development phase, and major production of soyasaponins A is mainly found at 5-week-old plant stage, while higher secretion of soya saponins B is mainly found at reproductive stages. The DDMP saponins were identified in trace levels during the development phases, even though they are important group of soyasaponin from tissues of root (Tsuno et al. 2018). These outcomes indicate mechanisms that control soya saponins production. Saponin compositions is highly variables in seeds of soybean based on seed tissues andvarieties, for example, hypocotyls, which ~2% of the total seed weight, contain more than 30% of saponin content and of which all are 'group A saponins' (Taniyama et al. 1988a; Shimoyamada et al. 1990; Shiraiwa et al. 1991a; Tsukamoto et al. 1993). Group A saponins are classified according to the variance in the C-22 sugar chains (Shiraiwa et al. 1991a, b; Mostafa et al. 2013a, b). The A saponin exhibits 2, 3, 4-tri-Oacetyl-b-D-xylopyranosyl(1 → 3)-a-L-arabino pyranosyl sugar, while Ab saponin contains a 2, 3, 4, 6-tetra-O-acetyl-b-D-glucopyranosyl(1 → 3)-a-L-arabino pyranosyl sugar and together of them have at the terminal position acetylated sugar (Fig. 3.3), which makes the unpleasant taste (Taniyama et al. 1988b; Okubo et al. 1992). On the other hand, A0-αgsaponins lack the acetylated terminalsugar at the C-22 position (Kikuchi et al. 1999; Takada et al. 2010). These discoveries have a significant useful role for breeding of soybean cultivar, used as human food, for example 'Kinusayaka', thease have astringent effects and lower bitter taste (Kato et al. 2007). Codominant alleles such as $Sg\text{-}1^b$and $Sg\text{-}1^a$, at a only locus designated *Sg-1* regulate the increase of saponins Ab andAa,; while a recessive allele, *sg-10* at the similar locules resulted in the biosynthesis of saponin A0-αg (Shiraiwa et al. 1990; Tsukamoto et al. 1993; Kikuchi et al. 1999; Takada et al. 2010). In addition, the biochemical studies of the allelic gene produces shown that $Sg\text{-}1^a$ and $Sg\text{-}1^b$ alleles code UDP-sugardependent glycosyl transferases, UGT73F2 and UGT73F4, which are involved in the catalyzing the adding of Xyl and Glc, to

Table 3.1 List of the ginseng saponins that have been isolated from different organs of Panax species

Saponin name	Organ	References
Floralginsenoside A Floralginsenoside B Floralginsenoside C Floralginsenoside D Floralginsenoside E Floralginsenoside F	Flower buds of *P. ginseng*	Yoshikawa et al. (2007a)
Floralginsenoside G Floralginsenoside H Floralginsenoside I Floralginsenoside J	Flower buds of *P. ginseng*	Nakamura et al. (2007a, b)
Floralginsenoside Ka Floralginsenoside Kb Floralginsenoside Kc	Flower buds of *P. ginseng*	Tung et al. (2010)
Floralginsenoside La Floralginsenoside Lb	Flower buds of *P. ginseng*	Nakamura et al. (2007a, b)
Floralginsenoside M Floralginsenoside N Floralginsenoside O Floralginsenoside P	Flower buds of *P. ginseng*	Yoshikawa et al. (2007b)
Floralginsenoside Ta Floralginsenoside Tb Floralginsenoside Tc	Flower buds of *P. ginseng*	Nguyen et al. (2010a, b)
Notoginsenoside FP1	Fruit pedicels *of P. notoginseng*	Wang et al. (2008)
Notoginsenoside FT2	*P. notoginseng* roots	Chen et al. (2006)
Notoginsenoside L Notoginsenoside M Notoginsenoside N	*P. notoginseng* roots	Yoshikawa et al. (2001)
Notoginsenoside O Notoginsenoside P Notoginsenoside Q	*P. notoginseng* flower buds	Yoshikawa et al. (2003)
Notoginsenoside Rw1	Flower buds of *P. notoginseng*	Cui et al. (2008)
Notoginsenoside R10	Roots of *P. ginseng*	Li et al. (2001)
Notoginsenoside S	Flower buds of *P. notoginseng*	Yoshikawa et al. (2003)
Notoginsenoside ST2 Notoginsenoside ST3	Steamed roots of *P. notoginseng*	Liao et al. (2008)
Notoginsenoside T1	Roots of *P. notoginseng*	Teng et al. (2004a, b)
Notoginsenoside T2	Roots of *P. notoginseng*	Teng et al. 2004a, 2004b
Notoginsenoside T3	Rhizomes of *P. notoginseng*	Cui et al. (2008)
Notoginsenoside T4 Notoginsenoside T5	Roots of *P. notoginseng*	Teng et al. (2004a, b)
Floralquinquenoside A Floralquinquenoside B Floralquinquenoside C Floralquinquenoside D Floralquinquenoside E	Flower buds of *P. quinquefolius*	Nakamura et al. (2007a, b)

(continued)

Table 3.1 (continued)

Saponin name	Organ	References
Ginsenoside I Ginsenoside II	Flower buds of *P. ginseng*	Qiu et al. (2001)
Ginsenoside Ki Ginsenoside Km	Steamed leaves *P. ginseng*	Tung et al. (2009)
Ginsenoside Rg6	Steamed leaves *P. ginseng*	Yang et al. (2000)
Ginsenoside Rg7	Leaves of *P. ginseng*	Dou et al. (2001)
Ginsenoside Rg8	Roots of *P. quinquefolius*	Dou et al. (2001)
Ginsenoside Rh5 Ginsenoside Rh6 Ginsenoside Rh7 Ginsenoside Rh8 Ginsenoside Rh9	Leaves of *P. ginseng*	Dou et al. (2001)
Ginsenoside Rk1 Ginsenoside Rk2 Ginsenoside Rk3	Processed roots of *P. ginseng*	Park et al. (2002a, b)
Ginsenoside Rs4 Ginsenoside Rs5 Ginsenoside Rs6 Ginsenoside Rs7	Steamed roots of *P. notoginseng*	Park et al. (2002a, b)
Ginsenoside Rz1	Steamed roots of *P. notoginseng*	Lee et al. (2009)
Quinquefoloside La	Leaves of *P. quinquefolius*	Jiang et al. (2008)
Quinquefoloside-Lb	*P. quinquefolius* leaves	Jiang et al. (2008)
Ginsenoside SL1 Ginsenoside SL2 Ginsenoside SL3 Ginsenoside ST1 Ginsenoside ST2	*P. ginseng* steamed leaves	Nguyen et al. (2010a, b)
Yesanchinoside A Yesanchinoside B Yesanchinoside C Yesanchinoside D Yesanchinoside E Yesanchinoside F	Underground part of *P. japonicus*	Zou et al. (2002a)
Yesanchinoside G Yesanchinoside H Yesanchinoside I	Underground part of *P. japonicus*	Zou et al. (2002b)
Quinquefoloside La	Leaves of *P. quinquefolius*	Jiang et al. (2008)
Quinquefoloside-Lb	Leaves of *P. quinquefolius*	Jiang et al. (2008)
Quinquefoloside Lc	Leaves of *P. quinquefolius*	Zhao et al. (2007)
Quinquenoside L16	Leaves and stems of *P. quinquefolius*	Chen et al. (2009)
Quinquenoside L1 Quinquenoside L2	Stems and leaves of *P. quinquefolius*	Wang et al. (2001a)
Quinquenoside L3	Stems and leaves of *P. quinquefolius*	Wang et al. (1998)
Quinquenoside L7	Stems and leaves of *P. quinquefolius*	Jiang et al. (2008)
Quinquenoside L9	Stems and leaves of *P. quinquefolius*	Wang et al. (2001b)

(continued)

Table 3.1 (continued)

Saponin name	Organ	References
Quinquenoside L10 Quinquenoside L14	Stems and leaves of *P. quinquefolius*	Chen et al. (2009)
Quinquenoside L17	Stems and leaves of *P. quinquefolius*	Li et al. (2009)
Floranotoginsenoside A Floranotoginsenoside B Floranotoginsenoside C Floranotoginsenoside D	*P. notoginseng* flowers	Wang et al. (2009)
Notopanaxoside A	Roots of *P. notoginseng*	Komakine et al. (2006)
Panaxadione	Seeds of *P. ginseng*	Sugimoto et al. (2009)

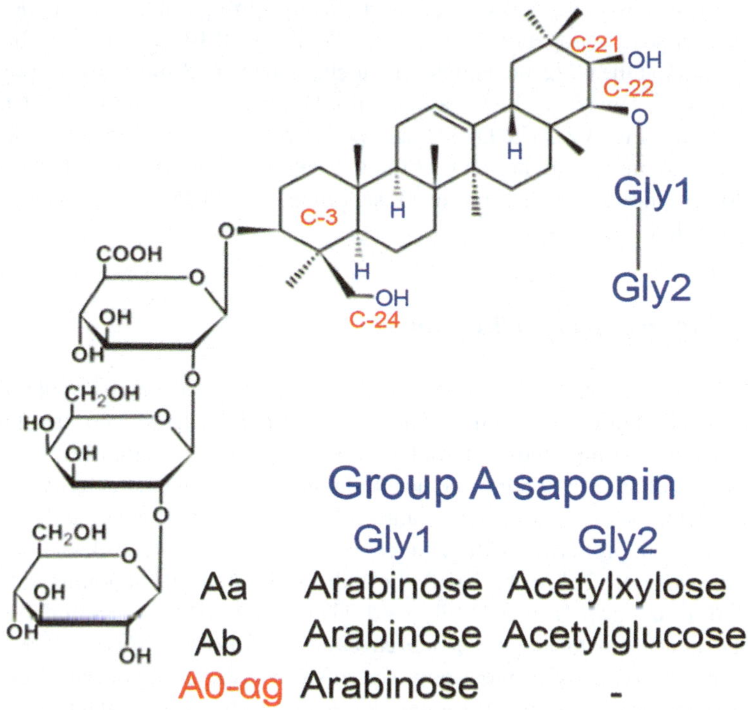

Fig. 3.3 Group A saponin in soybean and their differences at C-22 terminal sugar position

the Arabinose remains at the C-22 position (Sasama et al. 2012). A complete selection of wild germplasm of soybean resulted in the identification of a spontaneous mutant absent the capacity to synthesis soyasapogenol A, and the soya sapogenol A lack trait is existence unified into new soybean breeding lines (Sasama et al. 2010).

In addition, cultivar- and genotype-dependent variations in the content of saponin from soybean plants has been described, the composition of saponin 329 wild

(*Glycine soja*) and 800 cultivated (*Glycine max*) seeds of soybean presented a varied variety definite variances among the accessions. Similarly, large alterations of saponin composition in the seed of 3025 wild *G. soja* accessions in nine areas of Korea have been described, and composition of saponin ofseed hypocotyls was initially separated into 7 phenotypes, named as follow: Aa, Ab, AbBc, AaBc, AaBc +α Aa+α, and AbBc+α, where the major phenotypes remained in the following order AaBc, Aa, AaBc+α, and Aa+α (Panneerselvam et al. 2013). These variations in the saponins were accredited to variety exact expression of saponin biosynthetic genes that use soya sapogenol glycosides as substrates (Tsukamoto et al. 1993). In cultivated *G. max* 'B-type soya saponins', levels of represent one of the key metabolic pointers of connected salt tolerant and sensitive soybean verities (Wu et al. 2008). A-type saponins are bisdesmoside type that possessed 2 sugar chains at positions C-22 and C-3 of the hydroxyl groups of the aglycone, which nominated as soya sapogenol A (Fig. 3.2). While, DDMP saponins combined the DDMP moiety at the C-22 site and the sugar chain at the C-3 site of soyasapogenol B as the aglycone. However, the soyasapogenol B does not contain the C-21 position hydroxy group (Fig. 3.2). The DDMP and B saponins look to be generally dispersed with some variation in sugar chain at the C-3 site (Tsukamoto et al. 1993; Takada et al. 2012), indicating that group B saponins and DDMP may play a major biological role in soybean.

3.1.4 *Allium* Steroidal Saponins

Steroidal saponins are broadly distributed within monocot plants, containing the *Amaryllidaceae* family, where the *Allium* genus is being categorized. In addition to sulfur containg compounds, steroidal saponins are also significant bioactive metabolites that are measured to be accountable for the detected activity of several *Allium* species, such as cytotoxic, anti-inflammatory antifungal, and additional pharmacological properties (Mostafa et al. 2013a, b; Sobolewska et al. 2016; Abdelrahman et al. 2017a, b, 2019). In addition to steroidal saponins are widely found from *Amaryllidaceae* family and also some monocot families such as *Asparagaceae, Costaceae, Dioscoreaceae, Liliaceae, Melanthiaceae* and *Smilacaceae.* also dicotyledonous angiosperms such as Zygophyllaceae, Solanaceae Plantaginaceae and Fabaceae (Sobolewska et al. 2016). *Allium* plants suggest the utmost economically significant and a gorgeous basis of steroidal saponins with potential antifungal activity (Adao et al. 2011). Diverse *Allium* species, for example shallot (*A. cepa* L. Aggregatum group), garlic (*Allium sativum* L.), bulb onion (*Allium cepa* L.), chive (*Allium schoenoprasum* L.), and leek (*Allium porrum* L.), have been used in traditional medicines and food for a extended period (Fattorusso et al. 2000; Mostafa et al. 2013a, b). Steroidal saponins have been recognized so far in more than 50 various species of *Allium.* However, the initial reports on *Allium* saponins were describing the detection of alliogenin in *A. giganteum* bulbs (Khristulas et al. 1970) and diosgenin from *A. albidum* (Kereselidze et al. 1970). Twenty years later, Kravets et al. (1990) displayed the main chemical study of

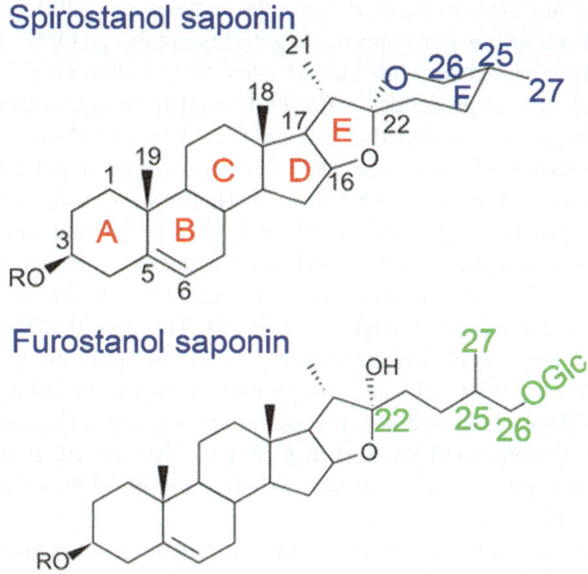

Fig. 3.4 A schematic view of furostanol and spirostanol saponin chemical structure

saponins from the *Allium* genus, and followed by further studies done by Lanzotti research group (Lanzotti 2005). Meanwhile, a large amount of new chemicals had been discovered from species of *Allium*. From the genus *Allium*, the steroidal saponins can be separated into three classes based on the sapogenin aglycone chemical structure: (1) furostanols, (2) spirostanols and (3) open chain cholestane saponins (Challinor and De Voss 2013). The moieties of sugar in *Allium* saponins are consisted of divided or lined chains prepared up most often of glucose, xylose rhamnose, galactose, and arabinose components (Abdelrahman et al. 2017a). Although, more research conducted has demonstrated the important of saponins as extraordinary antifungal compounds against various pathogens, few studies have measured describing the dissemination of total saponins inside the diverse organs from *Allium* plant species. A series of studies by Shigyo research groups (Mostafa et al. 2013a, b; Abdelrahman et al. 2014, 2017a, b) using different *Allium* species, indicated that the highest accumulation of total saponin can be originate from the basal stem and roots in comparison with leaves and bulbs. The functional point of assessment, the higher production of saponins from root and rootbasal stem compared with bulbs and leaves indicated that these are accountable for defensing host against various soil borne pathogens underground where the root basal stem is mostly found. Moreover, the quantitative difference of saponins and their transport in the root to the bulb, leaf, and/or flowers would be interlinked with environmental factors and progressive growth phase, as well as the interaction with plant pathogens and insects (Szakiel et al. 2011; Abdelrahman et al. 2014). In furostanol saponins whichever a *cis* or *trans* union among steroid ring A and B, or a double bond among C-5 and C-6 foremost to 5α, 5β or Δ5 series (Fig. 3.4). In addition, a double bond might also be positioned in chinenoside II, ascalonicoside B ceparoside C, (Sang et al. 2001; Fattorusso et al. 2002; Yuan et al. 2009). Furostanol saponins derived from *Allium* plant species, regularly contains an OMe or OH group at C-22 position. But, sapogenin with methyl ether C-22 is known to artifacts due to the methanol extraction procedures. *Allium* derived furostanol saponins are usually possessed a sugar chains bonded at C-3 and C-26 sites of bidesmodic glycoside units. However, unusual glycosylation at C-1 position with a galactose unit was also found in ascalonicosides A1/A2 saponin (Fattorusso et al. 2002). The most of furostanol saponin compounds usually exhibit an O-linked glucose residue linked at C-26 position. In saponin, including ceposides, persicoside C, and ascalonicosides A1/A2, a disaccharide chain was present at C-26 position (Fattorusso et al. 2002; Lanzotti 2012; Sadeghi et al. 2013). The steroid A/B ring junction from spirostane saponins is mainly detected in a *trans* (5α), or rarely from the *cis* (5β) fusion such as anzurogenin A/C and unsaturation is measured to be a relatively common feature (lilagenin, diosgenin, yuccagenin, ruscogenin, karatavigenin C; cepagenin). Though, a double bond situated at C25 was also described in the aglycones of saponins isolated from *A. macrostemon* and *A. ursinum* bulbs (Sobolewska et al. 2009; Cheng et al. 2013).

Recently, a phytochemical investigation on white bulb onion, resulted in the isolation and identification of the structure of three furostanol saponins metabolites called ceposides A, B and C (Lanzotti 2012). Ceposides A-C are recognized by a rare glycosylation of the hydroxyl group at C-1, and a similar skeletal character has been identified for the furostanol saponins called ascalonicosides and tropeosides, which were isolated from related onion species. Ceposide A-C, alone and in combinations were examined for their antifungal activity on ten different fungal species. The obtained results showed that all the three ceposides displayed a significant antagonistic activity against fungi based on their amount and the examined fungui. Additionally, a recent report by Abdelrahman et al. (2017a) identified a spirostanol saponin Alliospiroside A located in the TLC profiles of *A. fistulosum* with extra chromosome 2A in shallot. These results indicated that the modification of the *A. fistulosum* genome with shallot can improve the saponin profile with spirostanol-type saponin that showed potent antifungal activity against *Fusarium* pathogen. Furthermore, wild onion species *A. hirtifolium*, *A. elburzense*, *A. atroviolaceum*, and *A. minutiflorum* represent a potential genetic resources for steroidal saponins (Lanzotti 2012) that may add an additional insight for the Allium breeding to improve disease resistance.

References

Abdelrahman M, Hirata S, Ito S, Yamauchi N, Shigyo M (2014) Compartmentation and localization of bioactive metabolites in different organs of *Allium roylei*. Biosci Biotechnol Biochem 78:1112–1122

Abdelrahman M, El-Sayed M, Sato S, Hirakawa H, Ito S-I, Tanaka K, Mine Y, Sugiyama N, Suzuki Y, Yamauchi N, Shigyo M (2017a) RNA-sequencing-based transcriptome and biochemical analyses of steroidal saponin pathway in a complete set of *Allium fistulosum-A. cepa* monosomic addition lines. PLoS One 12:e0181784

Abdelrahman M, Mahmoud HYAH, El-Sayed M, Tanaka S, Tran LS (2017b) Isolation and characterization of Cepa2, a natural alliospiroside A, from shallot (*Allium cepa* L. Aggregatum group) with anticancer activity. Plant Physiol Biochem 116:167–173

Abdelrahman M, Hirata S, Sawada Y, Hirai MY, Sato S, Hirakawa H, Mine Y, Tanaka K, Shigyo M (2019) Widely targeted metabolome and transcriptome landscapes of *Allium fistulosum–A. cepa* chromosome addition lines revealed a flavonoid hot spot on chromosome 5A. Sci Rep 9:3541

Abe I, Rohmer M, Prestwich GC (1993) Enzymatic cyclization of squalene and oxidosqualene to sterols and triterpenes. Chem Rev 93:2189–2206

Adao CR, Da Silva BP, Parente JP (2011) A new steroidal saponin from *Allium ampeloprasum* var. porrum with antiinflammatory and gastroprotective effects. Phytochem Lett 4:306–310

Augustin JM, Kuzina V, Andersen SB, Bak S (2011) Molecular activities, biosynthesis and evolution of triterpenoid saponins. Phytochemistry 72:435–457

Baeg IH, So SH (2013) The world ginseng market and the ginseng (Korea). J Ginseng Res 37:1–7

Berhow MA, Wagner ED, Vaughn SF, Plewa MJ (2000) Characterization and antimutagenic activity of soybean saponins. Mutat Res 448:11–22

Berhow MA, Duval SM, Dobbins TA, Maynes J (2002) Analysis and quantitative determination of group B saponins in processed soybean products. Phytochem Anal 13:343–348

Challinor VL, De Voss JJ (2013) Open-chain steroidal glycosides, a diverse class of plant saponins. Nat Prod Rep 30:429–454

Chen JT, Li HZ, Wang D, Zhang YJ, Yang CR (2006) New dammarane monodesmosides from the acidic deglycosylation of notoginseng-leaf saponins. Helv Chim Acta 89:1442–1448

Chen J, Zhao R, Zeng YM, Meng H, Zuo WJ, Li X, Wang JH (2009) Three new triterpenoid saponins from the leaves and stems of *Panax quinquefolium*. J Asian Nat Prod Res 11:195–201

Cheng S-B, Wang Y, Zhang Y-F et al (2013) Steroidal saponins from *Allii macrostemonis* bulbs. Chin Tradit Herb Drug 44:1078–1081

Cibulski SP, Mourglia-Ettlin G, Teixeira TF, Quirici L, Roehe PM, Ferreira F, Silveira F (2016) Novel ISCOMs from *Quillaja brasiliensis* saponins induce mucosal and systemic antibody production, T-cell responses and improved antigen uptake. Vaccine 34:1162–1171

Cui XM, Jiang ZY, Zeng J, Zhou JM, Chen JJ, Zhang XM, Xu LS, Wang Q (2008) Two new dammarane triterpene glycosides from the rhizomes of *Panax notoginseng*. J Asian Nat Prod Res 10:845–849

de Costa F, Yendo AC, Cibulski SP, Fleck JD, Roehe PM, Spilki FR, Gosmann G, Fett-Neto AG (2016) Alternative inactivated poliovirus vaccines adjuvanted with *Quillaja brasiliensis* or Quil-a saponins are equally effective in inducing specific immune responses. PLoS One 9:e105374

de Faria JT, de Oliveira EB, Minim VPR, Minim LA (2017) Performance of Quillaja bark saponin and β-lactoglobulin mixtures on emulsion formation and stability. Food Hydrocoll 67:178–188

Dixit V, Tewari J, Obendorf SK (2010) Fungal growth inhibition of regenerated cellulose nanofibrous membranes containing Quillaja saponin. Arch Environ Contam Toxicol 59:417–423

Dou DQ, Chen YJ, Liang LH, Pang FG, Shimizu N, Takeda T (2001) Six new dammarane-type triterpene saponins from the leaves of *Panax ginseng*. Chem Pharm Bull (Tokyo) 49:442–446

Fattorusso E, Lanzotti V, Taglialatela-Scafati O, Di Rosa M, Lanaro A (2000) Cytotoxicsaponins from bulbs of *Allium porrum* L. J Agric Food Chem 48:3455–3462

Fattorusso E, Iorizzi M, Lanzotti V (2002) Chemical composition of shallot (*Allium ascalonicum* Hort.). J Agric Food Chem 50(20):5686–5690

Fleck JD, Kauffmann C, Spilki F, Lencina CL, Roehe PM, Gosmann G (2006) Adjuvant activity of *Quillaja brasiliensis* saponins on the immune responses to bovine herpesvirus type 1 in mice. Vaccine 24:7129–7134

Fleck JD, Betti AH, da Silva FP, Troian EA, Olivaro C, Ferreira F, Verza SG (2019) Saponins from *Quillaja saponaria* and *Quillaja brasiliensis*: particular chemical characteristics and biological activities. Molecules 24:171

Güçlü-Ustündağ O, Mazza G (2007) Saponins: properties, applications and processing. Crit Rev Food Sci Nutr 47:231–258

Guo S, Kenne L (2000) Structural studies of triterpenoid saponins with new acyl components from *Quillaja saponaria* Molina. Phytochemistry 55:419–428

Guo S, Lennart K, Lundgren LN, Rönnberg B, Sundquist BG (1998) Triterpenoid saponins from *Quillaja saponaria*. Phytochemistry 48:175–180

Han JY, Kwon YS, Yang DC, Jung YR, Choi YE (2006) Expression and RNA interference-induced silencing of the *dammarenediol synthase* gene in *Panax ginseng*. Plant Cell Physiol 47:1653–1662

Haralampidis K, Trojanowska M, Osbourn AE (2002) Biosynthesis of triterpenoid saponins in plants. In: Dutta NN et al (eds) History and trends in bioprocessing and biotransformation. Advances in biochemical engineering/biotechnology, vol 75. Springer, Berlin, Heidelberg, pp 31–49

Higuchi R, Tokimitsu Y, Komori T (1988) An acylated triterpenoid saponin from *Quillaja saponaria*. Phytochemistry 27:1165–1168

Holtshausen L, Chaves AV, Beauchemin KA, McGinn SM, McAllister TA, Odongo NE, Cheeke PR, Benchaar C (2009) Feeding saponin-containing *Yucca schidigera* and *Quillaja saponaria* to decrease enteric methane production in dairy cows. J Dairy Sci 92:2809–2821

Hong HD, Choi SY, Kim YC, Lee YC, Cho CW (2009) Rapid determination of ginsenosides Rb 1, Rf, and Rg 1 in Korean ginseng using HPLC. J Ginseng Res 33:8–12

Jacobsen NE, Fairbrother WJ, Kensil CR, Lim A, Wheeler DA, Powell MF (1996) Carbohydr Res 280:1–14

Jegal J, Jeong EJ, Yang MH (2019) A review of the different methods applied in ginsenoside extraction from *Panax ginseng* and *Panax quinquefolius* roots. Nat Product Commun 14:1–10

Jiang HP, Qiu YK, Cheng DR, Kang TG, Dou DQ (2008) Structure elucidation and complete NMR spectral assignments of two new dammarane-type tetraglycosides from *Panax quinquefolium*. Magn Reson Chem 46:786–790

Kahn RA, Durst F (2000) Function and evolution of plant cytochrome P450. Recent Adv Phytochem 34:151–189

Kaku T, Miyata T, Uruno T, Sako I, Kinoshita A (1975) Chemico-pharmacological studies on Saponins of *Panax Ginseng* C. A. Meyer. I. Chemical part. Arzneimittelforschung 25:343–347

Kato S, Yumoto S, Takada Y, Kono Y, Shimada S, Sakai T, Shimada H, Takahashi K, Adachi T, Tabuchi K, Kikuchi A (2007) A new soybean cultivar 'Kinusayaka' lacking three lipoxygenaseisozymes and group a acetyl saponin. Bull Natl Agric Res Cent Tohoku Reg 107:29–42

Kereselidze EV, Pkheidze TA, Kemertelidze EP (1970) Diosgenin from *Allium albidum*. Khim Prir Soedin 6(3):378

Khristulas FS, Gorovits MB, Luchanskaya VN et al (1970) A new steroid sapogenin from *Alliumgiganteum*. Khim Prir Soedin 6:489

Kikuchi A, Tsukamoto C, Tabuchi K, Adachi T, Okubo K (1999) Inheritance and characterization of a null allele for group Aacetyl saponins found in a mutant soybean (*Glycine max* (L.) Merrill). Breed Sci 49:167–171

Kim Y-J, Lee OR, Oh JY, Jang M-G, Yang D-C (2014) Functional analysis of 3-hydroxy-3-methylglutaryl coenzyme a reductase encoding genes in triterpene saponin-producing ginseng. Plant Physiol 165:373–387

Kitagawa I, Wang HK, Taniyama T, Yoshikawa AM (1998) Saponin and sapogenol A, soy sapogenol B, and soy sapogenol E oleanene sapogenols from soybean *Glycine max*. Structures of soy saponin I, soy saponin II, and soy saponin III. Chem Pharm Bull 36:153–161

Kite GC, Howes MJ, Simmonds MS (2004) Metabolomic analysis of saponins in crude extracts of *Quillaja saponaria* by liquid chromatography/mass spectrometry for product authentication. Rapid Commun Mass Spect 18:2859–2870

Kochan E, Szymańska G, Wielanek M, Wiktorowska-Owczarek A, Jóźwiak-Bębenista M, Grzegorczyk-Karolak I (2019) The content of triterpene saponins and phenolic compounds in American ginseng hairy root extracts and their antioxidant and cytotoxic properties. Plant Cell Tissue Org Cult 138:353–362

Komakine N, Okasaka M, Takaishi Y, Kawazoe K, Murakami K, Yamada Y (2006) New dammarane-type saponin from roots of *Panax notoginseng*. J Nat Med 60:137

Kravets SD, Vollerner YS, Gorovits MB et al (1990) Steroids of the spirostan and furostan series from plants of the genus *Allium*. Chem Nat Comp 26:359–373

Kushiro T, Ebizuka Y (2010) Triterpenes. In: Mander L, Liu HWB (eds) Comprehensive natural products II: chemistry and biology, vol 1. Elsevier, Oxford, pp 673–708

Kushiro T, Ohno Y, Shibuya M, Ebizuka Y (1997) In vitro conversion of 2,3-oxidosqualene into dammarenediol by *Panax ginseng* microsomes. Biol Pharm Bull 20:292–294

Kuzina V, Ekstrøm CT, Andersen SB, Nielsen JK, Olsen CE, Bak S (2009) Identification of defense compounds in *Barbarea vulgaris* against the herbivore *Phyllotreta nemorum* by an ecometabolomic approach. Plant Physiol 151:1977–1990

Kwon HW (2019) Inhibitory effects of ginsenoside Ro on clot retraction through suppressing PI3K/Akt signaling pathway in human platelets. Prev Nutr Food Sci 24:56–63

Lanzotti V (2005) Bioactive saponins from *Allium* and Aster plants. Phytochem Rev 4:95–110

Lanzotti V (2012) Bioactive polar natural compounds from garlic and onions. Phytochem Rev 11:179–196

Lee M, Shon HJ, Choi CS, Hung TM, Min BS, Bae K (2009) Ginsenosides from heat processed ginseng. Chem Pharm Bull 57:92–94

Lee MH, Han JY, Kim HJ, Kim YS, Huh GH, Choi YE (2011) Dammarenediol-II production confers TMV tolerance in transgenic tobacco expressing *Panax ginseng* dammarenediol-II synthase. Plant Cell Physiol 53:173–182

Lee DG, Lee J, Cho IH, Kim H-J, Lee S-W, Kim Y-O, Park C-G, Lee S (2017) Ginsenoside Rg12, a new dammarane-type triterpene saponin from *Panax ginseng* root. J Ginseng Res 41:531–533

Leshem Y, Levin I (1978) The effect of growing alfalfa on subsequent cotton plant development and on nitrate formation in peat soil. Plant Soil 50:323–328

Li HZ, Teng RW, Yang CR (2001) A novel hexanordammarane glycoside from the roots of *Panax nontoginseng*. Chin Chem Lett 12:59–62

Li GY, Zeng YM, Meng H, Li X, Wang JH (2009) A new triterpenoid saponin from the leaves and stems of *Panax quinquefolium* L. Chin Chem Lett 20:1207–1210

Liao PY, Wang D, Zhang YJ, Yang CR (2008) Dammarane-type glycosides from steamed notogensing. J Agric Food 56:1751–1756

Lu JM, Jiang J, Jamaluddin MS, Liang Z, Yao Q, Chen C (2019) Ginsenoside Rb1 blocks ritonavir-induced oxidative stress and ENOS downregulation through activation of estrogen receptor-beta and upregulation of SOD in human endothelial cells. Int J Mol Sci 20:294

Luo HM, Sun C, Sun YZ, Wu Q, Li Y, Song JY, Niu YY, Cheng X, Xu HX, Li CY et al (2011) Analysis of the transcriptome of *Panax notoginseng* root uncovers putative triterpene saponin–biosynthetic genes and genetic markers. BMC Genome 12:S5

MacDonald RS, Guo JY, Copeland J, Browning JD, Sleper JD, Rottinghaus GE, Berhow MA (2005) Environmental influences on isoflavones and saponins in soybeans and their role in colon cancer. J Nutr 135:1239–1242

Maier C, Conrad J, Carle R, Weiss J, Schweiggert RM (2015) Phenolic constituents in commercial aqueous Quillaja (*Quillaja saponaria* Molina) wood extracts. J Agric Food Chem 63:1756–1762

Mohanan P, Subramaniyam S, Mathiyalagan R, Yang DC (2018) Molecular signaling of Ginsenosides Rb1, Rg1, and Rg3 and their mode of actions. J Ginseng Res 42:123–132

Moses T, Papadopoulou KK, Osbourn A (2014) Metabolic and functional diversity of saponins, biosynthetic intermediates and semi-synthetic derivatives. Crit Rev Biochem Mol Biol 49:439–462

Mostafa A, Jogaiah S, El-Sayed M, Ito S-I et al (2013a) Aginoside saponin, a potent antifungal compound, and secondary metabolite analyses from *Allium nigrum* L. Phytochem Lett 6:274–280

Mostafa A, Sudisha J, El-Sayed M, Ito S-I, Ikeda T, Yamauchi N, Shigyo M (2013b) Aginoside saponin, a potent antifungal compound, and secondary metabolite analyses from *Allium nigrum* L. Phytochem Lett 6:274–280

Nakamura S, Sugimoto S, Matsuda H, Yoshikawa M (2007a) Medicinal flowers. XVII. New dammarane-type triterpene glycosides from flower buds of American ginseng, *Panax quinquefolium* L. Chem Pharm Bull (Tokyo) 55:1342–1348

Nakamura S, Sugimoto S, Matsuda H, Yoshikawa M (2007b) Structures of dammarane-type triterpene triglycosides from the flower buds of *Panax ginseng*. Heterocycles 71:577–588

Nguyen HT, Song GY, Kang HK, Kim YH (2010a) New dammarane saponins from the steamed ginseng leaves. Bull Kor Chem Soc 31:2094–2096

Nguyen HT, Song GY, Kim JA, Hyun JH, Kang HK, Kim YH (2010b) Dammarane-type saponins from the flower buds of *Panax ginseng* and their effects on human leukemia cells. Bioorg Med Chem Lett 20:309–314

Niu YY, Luo HM, Sun C, Yang T-J, Dong L, Huang LF, Chen SL (2014) Expression profiling of the triterpene saponin biosynthesis genes *FPS*, *SS*, *SE*, and *DS* in the medicinal plant *Panax notoginseng*. Gene 533:295–303

Nord LI, Kenne L (2000) Novel acetylated triterpenoid saponins in a chromatographic fraction from *Quillaja saponaria* Molina. Carbohydr Res 329:817–829

Nyberg NT, Baumann H, Kenne L (2003) Solid-phase extraction NMR studies of chromatographic fractions of saponins from *Quillaja saponaria*. Anal Chem 75:268–274

Okubo K, Iijima M, Kobayashi Y, Yoshikoshi M, Uchida T, Kudou S (1992) Components responsible for the undesirable taste of soybean seeds. Biosci Biotechnol Biochem 56:99–103

Oleszek W (1993) Allelopathic potentials of alfalfa (*Medicago sativa*) saponins: their relation to antifungal and hemolytic activities. J Chem Ecol 19:1063–1074

Osbourn AE (2003) Saponins in cereals. Phytochemistry 62:1–4

Panneerselvam K, Tsukamoto C, Honda N et al (2013) Saponin polymorphism in the Korean wild soybean (*Glycine soja* Sieb. and Zucc.). Plant Breed 132:121–126

Parente JP, Da Silva BP (2009) Bioactive complex triterpenoid saponins from the Leguminosae family. Nat Prod Commun 4:143–155

Park IH, Han SB, Kim JM, Piao LZ, Kwon SW, Kim NY, Kang TL, Park MK, Park JH (2002a) Four new acetylated ginsenosides from processed ginseng (sun ginseng). Arch Pharm Res 25:837–841

Park IH, Kim NY, Han SB, Kim JM, Kwon SW, Kim HJ, Park MK, Park JH (2002b) Three new dammarane glycosides from heat processed ginseng. Arch Pharm Res 25:428–432

Patel SS, Savjani JK (2015) Systematic review of plant steroids as potential anti-inflammatory agents: current status and future perspectives. Phytopharmacology 4:121–125

Pen B, Sar C, Mwenya B, Kuwaki K, Morikawa R, Takahashi J (2006) Effects of *Yucca schidigera* and *Quillaja saponaria* extracts on in vitro ruminal fermentation and methane emission. Anim Feed Sci Technol 129:175–186

Qi LW, Wang C-Z, Yuan C-S (2011) Isolation and analysis of ginseng: advances and challenges. Nat Prod Rep 28:467–495

Qiu F, Ma ZZ, Xu SX, Yao XS, Che CT, Chen YJ (2001) A pair of 24-hydroperoxyl epimeric dammarane saponins from flower-buds of *Panax ginseng*. J Asian Nat Prod Res 3:235–240

Reichert CL, Salminen H, Bönisch GB, Schäfer C, Weissa J (2019) Concentration effect of *Quillaja* saponin - co-surfactant mixtures on emulsifying properties. Colloid Interface Sci 519:71–80

Roner MR, Sprayberry J, Spinks M, Dhanji S (2007) Antiviral activity obtained from aqueous extracts of the Chilean soapbark tree (*Quillaja saponaria* Molina). J Gen Virol 88:275–285

Roner MR, Tam KI, Kiesling-Barrager M (2010) Prevention of rotavirus infections in vitro with aqueous extracts of *Quillaja Saponaria* Molina. Future Med Chem 2:1083–1097

Sadeghi M, Zolfaghari B, Senatore M, Lanzotti V (2013) Spirostane, furostane and cholestane saponins from Persian leek with antifungal activity. Food Chem 141:1512–1521

San Martín R, Briones R (1999) Industrial uses and sustainable supply of *Quillaja saponaria* (Rosaceae) saponins. Econ Bot 53:302–311

Sang S, Mao S, Lao A, Chen Z, Ho CT (2001) Four new steroidal saponins from the seeds of *Allium tuberosum*. J Agric Food Chem 49:1475–1478

Sasama H, Takada Y, Ishimoto M, Kitamura K, Tsukamoto C (2010) Estimation of the mutation site of a soyasapogenol A-deficient soybean [*Glycine max* (L.) Merr.] by LC-MS/MS profile analysis. In: Cadwallader KR, Chang S (eds) Chemistry, texture, and flavor of soy, vol 1059. American Chemical Society, New York, pp 91–102

Sasama T, Ono E, Takagi K, Takada Y, Horikawa M, Nakamoto Y et al (2012) The *Sg-1* glycosyltransferase locus regulates structural diversity of triterpenoid saponins of soybean. Plant Cell 24:2123–2138

Sawai S, Saito K (2011) Triterpenoid biosynthesis and engineering in plants. Front Plant Sci 30. https://doi.org/10.3389/fpls.2011.00025

Shi J, Arunasalam K, Yeung D et al (2004) Saponins from edible legumes: chemistry, processing, and health benefits. J Med Food 7:67–78

Shi Z-Y, Zeng J-Z, Wong AST (2019) Chemical structures and pharmacological profiles of ginseng saponins. Molecules 24:2443

Shimoyamada M, Kudou S, Okubo K, Yamauchi F, Harada K (1990) Distribution of saponin constituents in some varieties of soybean plant. Agric Biol Chem 54:77–81

Shin KC, Oh DK (2016) Classification of glycosidases that hydrolyze the specific positions and types of sugar moieties in ginsenosides. Crit Rev Biotechnol 36:1036–1049

Shin B-K, Kwon SW, Park JH (2015) Chemical diversity of ginseng saponins from *Panax ginseng*. J Ginseng Res 39:287–298

Shiraiwa M, Yamauchi F, Harada K, Okubo K (1990) Inheritance of "group A saponin" in soybean seed. Agric Biol Chem 54:1347–1352

Shiraiwa S, Harada K, Okubo K (1991a) Composition and content of saponins in soybean seed according to variety, cultivation year and maturity. Agric Biol Chem 55:323–331

Shiraiwa M, Harada K, Okubo K (1991b) Composition andstructure of "group B saponin" in soybean seed. Agric Biol Chem 55:911–917

Sobolewska D, Janeczko Z, Podolak I et al (2009) Densitometric analysis of diosgenin in methanolic extracts of *Allium ursinum* collected at different times during plant development. J Planar Chromatogr 22:305–307

Sobolewska D, Michalska K, Podolak I, Grabowska K (2016) Steroidal saponins from the genus *Allium*. Phytochem Rev 15:1–35

Sparg SG, Light ME, van Staden J (2004) Biological activities and distribution of plant saponins. J Ethnopharmacol 94:219–243

Sugimoto S, Nakamura S, Matsuda H, Kitagawa N, Yoshikawa M (2009) Chemical constituents from seeds of *Panax ginseng*: structure of new dammarane-type triterpene ketone, panaxadione, and HPLC comparisons of seeds and flesh. Chem Pharm Bull 57:283–287

Sugiyama A (2019) The soybean rhizosphere: metabolites, microbes, and beyond a review. J Adv Res 19:67–73

Sun H-X, Xie Y, Ye Y-P (2009) Advances in saponin-based adjuvants. Vaccine 27:1787–1796

Szakiel A, Pączkowski C, Henry M (2011) Influence of environmental abiotic factors on the content of saponins in plants. Phytochem Rev 10:471–491

Takada Y, Sayama T, Kikuchi A, Kato S, Tatsuzaki N, Nakamoto Y, Suzuki A, Tsukamoto C, Ishimoto M (2010) Genetic analysis of variation in sugar chain composition atthe C-22 position of group A saponins in soybean, *Glycine max* (L.). Merrill. Breed Sci 60:3–8

Takada Y, Tayama I, Sayama T, Sasama H, Sarut M, Kikuchi A, Ishimoto M, Tsukamoto C (2012) Genetic analysis of variations in the sugar chain composition at the C-3 position of soybean seed saponins. Breed Sci 61:639–645

Tam KI, Roner MR (2011) Characterization of in vivo anti-rotavirus activities of saponin extracts from *Quillaja saponaria* Molina. Antivir Res 90:231–241

Taniyama T, Yoshikawa M, Kitagawa I (1988a) Saponinand sapogenol. XLIV. Soyasaponin composition in soybeans of variousorigins and soyasaponin content in various organs of soybean. Structure of soyasaponin V from soybean hypocotyl. Yakugaku Zasshi 108:562–571

Taniyama T, Nagahama Y, Yoshikawa M, Kitagawa I (1988b) Saponin and sapogenol. XLIII. Acetyl-soyasaponins A4, A5, and A6, new astringent bisdesmosides of soyasapogenol A, from Japanese soybean, the seeds of *Glycine max* MERRILL. Chem Pharm Bull (Tokyo) 36:2829–2839

Tansakul P, Shibuya M, Kushiro T, Ebizuka Y (2006) Dammarenediol-II synthase, the first dedicated enzyme for ginsenoside biosynthesis, in *Panax ginseng*. FEBS Lett 580:5143–5149

Teng RW, Ang C, McManus D, Armstrong D, Mau S, Bacic A (2004a) Regioselective acylation of ginsenosides by Novozyme 435 to generate molecular diversity. Helv Chim Acta 87:1860–1872

Teng RW, Li HZ, Wang DZ, Yang CR (2004b) Hydrolytic reaction of plant extracts to generate molecular diversity: new dammarane glycosides from the mild acid hydrolysate of root saponins of *Panax notoginseng*. Helv Chim Acta 87:1270–1278

Tsukamoto C, Kikuchi A, Harada K et al (1993) Genetic and chemical polymorphisms of saponins in soybean seed. Phytochemistry 34:1351–1356

Tsuno Y, Fujimatsu T, Endo K, Sugiyama A, Yazaki K (2018) Soyasaponins: a new class of root exudates in soybean (*Glycine max*). Plant Cell Physiol 59:366–375

Tung NH, Song GY, Park YJ, Kim YH (2009) Two new dammarane-type saponins from the leaves of *Panax ginseng*. Chem Pharm Bull (Tokyo) 57:1412–1414

Tung NH, Song GY, Nhiem NX, Ding Y, Tai BH, Jin LG, Lim CM, Hyun JW, Park CJ, Kang HK, Kim YH (2010) Dammarane-type saponins from the flower buds of *Panax ginseng* and their intracellular radical scavenging capacity. J Agric Food Chem 58:868–874

Turner TR, Ramakrishnan K, Walshaw J et al (2013) Comparative meta-transcriptomics reveals kingdom level changes in the rhizosphere microbiome of plants. ISME J 7:2248–2258

Wang J, Li W, Li X (1998) A new saponin from the leaves and stems of *Panax quinquefolium* L. collected in Canada. J Asian Nat Prod Res 1:93–97

Wang JH, Li W, Sha Y, Tezuka Y, Kadota S, Li X (2001a) Triterpenoid saponins from leaves and stems of *Panax quinquefolium* L. J Asian Nat Prod Res 3:123–130

Wang JH, Sha Y, Tezuka Y, Kadota S, Li X (2001b) Quinquenoside L9 from leaves and stems of *Panax quinquefolium* L. J Asian Nat Prod Res 3:293–297

Wang XY, Wang D, Ma XX, Zhang Y-J, Yang C-R (2008) Two new dammarane-type bisdesmosides from the fruit pedicels of *Panax notoginseng*. Helv Chim Acta 91:60–66

Wang J-R, Yamasaki Y, Tanaka T, Kouno I, Jiang Z-H (2009) Dammarane-type triterpene saponins from the flowers of *Panax notoginseng*. Molecules 14:2087–2094

Wojciechowski K (2013) Surface activity of saponin from *Quillaja* bark at the air/water and oil/water interfaces. Colloids Surf B Biointerfaces 108:95–102

Wong AST, Che CM, Leung KW (2015) Recent advances in ginseng as cancer therapeutics: a functional and mechanistic overview. Nat Prod Rep 32:256–272

Wu W, Zhang Q, Zhu Y et al (2008) Comparative metabolic profiling reveals secondary metabolites correlated with soybean salt tolerance. J Agric Food Chem 56:11132–11138

Xu J, Chu Y, Liao B, Xiao S, Yin Q, Bai R, Su H, Dong L, Li X, Qian J et al (2017) *Panax Ginseng* genome examination for ginsenoside biosynthesis. Gigascience 6:1–15

Yang XW, Li LY, Tian JM, Zhang ZW, Ye JM, Gu WF (2000) Ginsenoside-Rg6, a novel triterpenoid saponin from the stem-leaves of *Panax ginseng* CAMey. Chin Chem Lett 11:909–912

Yang J-L, Hu Z-F, Zhang T-T, Gu A-D, Gong T, Zhu P (2018) Progress on the studies of the key enzymes of ginsenoside biosynthesis. Molecules 23:589

Yoshikawa M, Morikawa T, Yashiro K, Murakami T, Matsuda H (2001) Bioactive saponins and glycosides. XIX. Notoginseng (3): immunological adjuvant activity of notoginsenosides and related saponins: structures of notoginsenosides-L, -M, and -N from the roots of *Panax notoginseng* (Burk.) F.H. Chen. Chem Pharm Bull (Tokyo) 49:1452–1456

Yoshikawa M, Morikawa T, Kashima Y, Ninomiya K, Matsuda H (2003) Structures of new dammarane-type triterpene saponins from the flower buds of *Panax notoginseng* and hepatoprotective effects of principal ginseng saponins. J Nat Prod Res 66:922–927

Yoshikawa M, Sugimoto S, Nakamura S, Sakumae H, Matsuda H (2007a) Medicinal flowers. XVI. New dammarane-type triterpene tetraglycosides and gastroprotective principles from flower buds of *Panax ginseng*. Chem Pharm Bull 55:1034–1038

Yoshikawa M, Sugimoto S, Nakam S, Matsuda H (2007b) Medicinal flowers. XI. Structures of new dammarane-type triterpene diglycosides with hydroperoxide group from flower buds of *Panax ginseng*. Chem Pharm Bull (Tokyo) 55:571–576

Yuan L, Ji TF, Li CJ, Wang AG, Yang JB, Su YL (2009) Two new steroidal saponins from the seeds of *Allium cepa* L. J Asian Nat Prod Res 11:213–218

Zhao Y, Wang W, Han L, Rayburn ER, Hill DL, Wang H, Zhang R (2007) Isolation, structural determination, and evaluation of the biological activity of 20(S)-25-methoxyl-dammarane-3beta, 12beta, 20-triol [20(S)-25-OCH3-PPD], a novel natural product from *Panax notoginseng*. Med Chem 3:51–60

Zhu S, Zou K, Fushimi H, Cai S, Komatsu K (2004) Comparative study on triterpene saponins of ginseng drugs. Planta Med 70:666–677

Zou K, Zhu S, Meselhy Tohda M, Cai S, Komatsu K (2002a) Dammarane-type saponins from *Panax japonicus* and their neurite outgrowth activity in SK-N-SH cells. J Nat Prod 65:1288–1292

Zou K, Zhu S, Tohda C, Cai S, Komatsu K (2002b) Dammarane-type Triterpene Saponins from *Panax japonicas*. J Nat Prod 65:346–351

Saponins Versus Plant Fungal Pathogens

4

Abstract

Saponins, a group of phytoanticipins are recognized as the first biochemical barriers against wide range of fungal pathogens. Although the detailed mechanisms of saponin antifungal mode of action is not well established, it is believed that saponin aglycone-sugar structure forms complex with the pathogen sterols in the cell membrane, leading to lose of the cell membrane and pore formation and consequently loss of membrane integrity. In this chapter we will discussed different saponin compounds and their mode of action against wide range of phytopathogens.

4.1 Introduction

Plants synthesis a vast array of bioactive metabolites, and many of which can suppress the growth of plant pathogens *in vitro*. These bioactive metabolites can be synthesized during normal plant growth and development stages, and usually called phytoanticipins (Schönbeck and Schlösser 1976; VanEtten et al. 1994; Mostafa et al. 2013; da Cruz Cabral et al. 2013; Matušinský et al. 2015); or alternatively synthesized only when plants are subjected to stress or pathogen attack and in that case they called phytoalexins (Papadopoulou et al. 1999; Abdelrahman et al. 2014). Thus, the antimicrobial compounds incorporate a diverse array of various classes of bioactive compounds, including saponins, isoflavonoids, alkaloids, phenolics, terpenoids, cyclic hydroxamic acids, sulfur-containing compounds and others. Since many saponins (glycosylated triterpenoid or steroid molecules) showed potential antifungal characteristics, these molecules have been implicated a chemical barriers against wide range of fungal pathogens, however there is no reports regarding the antibacterial activity of saponin compounds (Mostafa et al. 2013; Abdelrahman et al. 2014). Saponin compounds encompass various family of triterpenoids and steroids according to their chemical structure

M. Abdelrahman, S. Jogaiah, *Bioactive Molecules in Plant Defense*,
https://doi.org/10.1007/978-3-030-61149-1_4

(Podolak et al. 2010; Moses et al. 2014). Saponins display amphiphilic characters due to the presence of a lipophilic triterpene aglycon (sapogenin) attached with hydrophilic sugar chains, which both facilities saponin binding activity into the fungal membrane. The antifungal properties of the saponin molecules were attributed to their core aglycone moieties as well as number and structure of saccharide units in their sugar chain (Yang et al. 2006; Abdelrahman et al. 2017). The antifungal properties of saponin molecules are mainly interlinked with their capability to bind with the sterols component of fungi cell membrane, causing membrane perturbation and subsequently leakage of cell contents (Yang et al. 2006; Podolak et al. 2010; Sreij et al. 2019). However, the specific role of saponin activity in plant–pathogen interaction is still not properly known (Trdá et al. 2019). Some fungi species have the ability to protect themselves from saponin toxicity due to their ability to secret saponin-detoxifying enzymes, while other have intrinsic resistance due to the special structure of their cell membrane. The antifungal activity of saponin molecules has been well documents, especially their activity against phytopathogens of crop (Barile et al. 2007; Teshima et al. 2013; Abdelrahman et al. 2017). However, only few studies have compared the antifungal activity against phytopathogenic fungi with respect to commercial fungicides (Saniewska et al. 2006; Porsche et al. 2018). In addition, Yu et al. (2013) reported several biochemical changes that could be associated with the probable mechanisms of the antimicrobial activity of saponin compounds, including reduce catalase activity, decrease glucose utilization rate and protein content in microbial pathogens. However, the exact molecular function of the saponin compound still not fully understood in plant-pathogen interaction system.

4.2 Steroidal Saponins Isolated from *Allium* Crops and Their Antifungal Properties

Barile et al. (2007) successfully isolated three saponin compounds, named minutoside A, B and C, in addition to two previously reported sapogenins, named neoagigenin, and alliogenin from the bulbs of *Allium minutiflorum*. Based on 2D nuclear magnetic resonance (NMR) spectroscopy, the structures of the three newly isolated saponin compounds were identified as follow: (25R)-furost-2alpha,3beta,6beta,22alpha,26-pentaol 3-O-[beta-D-xylopyranosyl-(1→3)-O-beta-D-glucopyranosyl-(1→4)-O-beta-D-galactopyranosyl] 26-O-beta-D-glucopyranoside (minutoside A), (25S)-spirostan-2alpha,3beta,6beta-triol 3-O-beta-D-xylopyranosyl-(1→3)-O-beta-D-glucopyranosyl-(1→4)-O-beta-D-galactopyranoside (minutoside B), and (25R)-furost-2alpha,3beta,5alpha,6beta,22alpha,26-esaol 3-O-[beta-D-xylopyranosyl-(1→3)-O-beta-D-glucopyranosyl-(1→4)-O-beta-D-galactopyranosyl] 26-O-beta-D-glucopyranoside (minutoside C). All the three new saponin compounds showed a remarkable antifungal activity depending on their concentration and among which, minutoside B displayed the highest activity, while minutoside A showed the lowest antifungal activity (Barile et al. 2007). However, no antibacterial activity was found in the tested saponin compounds. This result indicated the significant role of spirostanol-type aglycone for the antifungal activity. Many phenotypic alterations were observed in the tested fungi, including changes in the sporulation rate and swelling of the fungi hypha (Barile et al. 2007). Among tested fungi, *Alternaria alternate, A. porri, Botrytis cinerea, Fusarium oxysporum, F. oxysporum f. sp. lycopersici, F. solani, Pythium ultimum, Rhizoctonia solani, Trichoderma harzianum* P1 and *T. harzianum* T39. Among which, the two strains of *T. harzianum* P1 and T39 were more sensitive than other fungal pathogens to minutosides B and C saponins, which was consistence with previous report regarding the high sensitivity of *Trichoderma* spp. to saponins isolated from *Panax quinquefolius* and *Medicago sativa* (Zimmer et al. 1967; Nicol et al. 2002). On the other hand *Pythium ultimum*, was more resistant to most of the examined saponins, and this could be partially explained by lack of sterols in the membrane of oomycetes *P. ultimum* and thus saponins could not bind and perform its activity (Barile et al. 2007). In another study by Teshima et al. (2013) successfully isolated two known saponin compounds named alliospiroside A and B from the root of shallot plants. Alliospiroside A showed potent antifungal activity against wide range of tested pathogens, including *Alternaria alternate, A. solani, A. tenuissima, B. cinerea, B. squamosa, Colletotrichum acutatum, C. gloeosporioides, C. graminicola, Curvularia lunata, Epicoccum nigrum, Fusarium oxysporum* f. sp. batatas, *F. oxysporum* f. sp. cepae, *F. solani, F. proliferatum, F. verticillioides, Magnaporthe oryzae, Sclerotium cepivorum* and *Thanatephorus cucumeris* (Teshima et al. 2013). Both alliospiroside A and B inhibited the growth of all tested fungi pathogen, tested, and among which, *Colletotrichum* spp. were the most sensitive to the saponins, whereas the *Fusarium* spp. were more resistant to the saponins. In addition, the authors suggested that the induction of reactive oxygen species (ROS), was one of the mode of action involved in the fungicidal action of saponin, which was evident by the rapid accumulation of ROS in *C. gloeosporioides* cells treated with alliospiroside A followed by DHR staining, which was also correlated with the extent of dead fungal cells stained with propidium iodide or Evans blue dye. In agreement with Teshima et al. (2013), Alliospiroside A saponin compound was successfully isolate and identify in *A. fistulosum* (FF) with extra chromosome 2A from shallot (FF2A) (Abdelrahman et al. 2017). The saponin TLC profile showed a characteristic saponin band on the shallot (AA) and FF2A; however, this saponin spot was lacking in other monosomic addition lines (MALs) and FF profile (Abdelrahman et al. 2017). Furthermore, two furostanol saponin compounds were observed in the AA, FF1A, and FF2A saponin profile relative to FF and other MALs. The total content of saponin was highly rich in the root of AA, FF1A and FF2A relative to other MALs and FF (Abdelrahman et al. 2017). The authors suggested that the genes set involved in saponin biosynthesis could be present on the chromosome 2A of shallot, and these saponin-related gene(s) are contributed to the characteristic saponin compound observed (Abdelrahman et al. 2017). The structure of the isolated pure compound was interpreted by 600 MHz NMR and both 1H NMR and 13C NMR data of the pure compound was identical to the spirostnaol-type saponin named Alliospiroside A, with chemical structure [[(25S)-3β-hydroxyspirost-5-en-

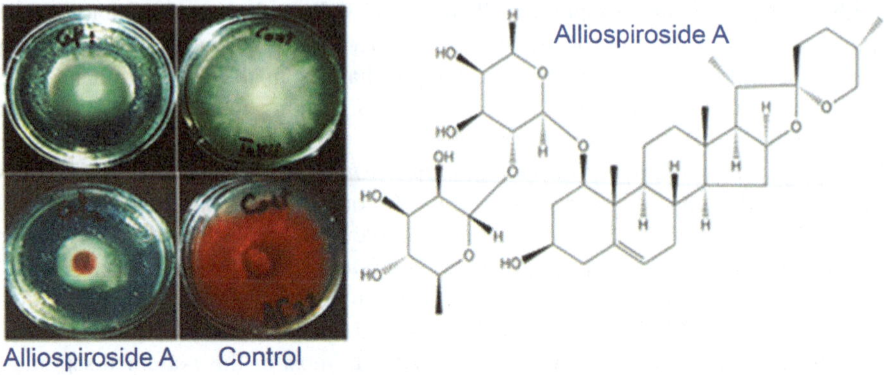

Fig. 4.1 Chemical structure of Alliospiroside A isolated from *Allium fistulosum* with additional chromosome 2A from shallot, and its antifungal activities against some fusarium pathogen

1β-yl] 2-O-(6-deoxy-α-L-mannopyranosyl)-α-L-arabinopyranoside] and a molecular formula of C38H60O12 (Fig. 4.1). To evaluate the biological role of spirostanol saponin Alliospiroside A in disease resistance against *Fusarium* pathogens, first the crude saponin extracts derived from FF, MALs and AA dry roots were tested against two *F. oxysporum* f. sp. cepa strains named AF22 and TA using agar diffusion method (Fig. 4.1). The obtained results showed highest fungal growth inhibition percentage by FF2A-root saponin against the two *F. oxysporum* f. sp. cepa strains (Fig. 4.1). Furthermore, Alliospiroside A antifungal activity was much superior than the furostanol saponin fraction with respective pathogens.

In another study carried out by Italian group was able to isolate three saponin compounds, namely ceposide A, B, and C from the bulbs of white onion (*A. cepa*) (Lanzotti et al. 2012b). The chemical structure of the three saponin compounds was conducted by 2D NMR spectroscopy and mass spectrometry (MS). The structures of the saponin compounds were showed as follow, (1) ceposide A: (25R)-furost-5(6)-en-1β,3β,22α,26-tetraol 1-O-β-D-xylopyranosyl 26-O-α-D-rhamnoyranosyl-(1→2)-O-β-D-galactopyranoside, (2) ceposide B: (25R)-furost-5(6)-en-1β,3β,22α,26-tetraol 1-O-β-D-xylopyranosyl 26-O-α-D-rhamnoyranosyl-(1→2)-O-β-D-glucopyranoside, and (3) ceposide C: (25R)-furost-5(6)-en-1β,3β,22α,26-tetraol 1-O-β-D-galactopyranosyl 26-O-α-D-rhamnoyranosyl-(1→2)-O-β-D-galactopyranoside. The isolated saponin compounds, alone or in combinations, were tested for their antifungal properties against several fungal species, including *Aspergillus niger*, *F. oxysporum* f. sp. lycopersici, *A. alternata*, *B. cinerea*, *Phomopsis* sp., *Mucor* sp., *Sclerotium cepivorum*, *Rhizoctonia solani*, *T. harzianum*, and *T. atroviride* (Lanzotti et al. 2012b). The data results showed that ceposide B exhibited the highest antifungal activity, while ceposide C was the lowest. In addition, both *B. cinerea* and *T. atroviride* were very sensitive to the saponin application, on the other hand *F. oxysporum* f. sp. lycopersici, *R. solani* and *S. cepivorum* were highly resistant to the saponins (Lanzotti et al. 2012a, b). The early study by Mostafa et al. (2013) using phytochemical investigations of *A. nigrum* root extract resulted in the isolation of a spirostane-type glycoside named aginoside. The chemical structure of aginoside saponin was elucidated by 2D NMR, FABMS and HR-ESI-MS analysis, and was identified as 25(R,S)-5α-spirostan-2α,3β,6-β-triol-3-O-β-d-glucopyranosyl-(1→2)-O-[β-d-xylopyranosyl-(1→3)]-O-β-d-glucopyranosyl-(1→4)-β-d-galactopyranoside. Aginoside sapirostanol saponin was able to strongly inhibit *F. oxysporum* and *C. gloeosporioides* phytopathogens. The high ability of aginoisde saponin even against Fusarium pathogen suggested that a cross breeding strategy to induce aginoisde saponin in the cultivated *Allium* might be potential strategy to induce disease resistance against fusarium disease. Likewise, the phytochemical analysis of garlic (*A. sativum*) buld extracts, enabled the isolation and identification of several furostanol saponin compounds named; voghieroside A1/A2 and B1/B2, C1/C2, D1/D2 and E1/E2 based on the rare agapanthagenin aglycone; agigenin aglycone; and gitogenin aglycone, respectively (Lanzotti et al. 2012a). Moreover, two known spirostanol saponins, gitogenin 3-O-tetrasaccharide and agigenin 3-O-trisaccharide were detected. The antifungal activity of the isolated saponin compounds was evaluated against two fungal species *T. harzianum* and *B. cinerea* on dose dependent concentration. All saponin compounds showed potential antifungal activity against *T. harzianum*, while *B. cinerea* was slightly more resistant than *T. harzianum*. In general, voghieroside A exhibited the lowest antifungal activity compared with other isolated saponin compounds (Lanzotti et al. 2012a). The list of some isolated saponin compounds from different *Allium* species have been listed in the following Table 4.1.

4.3 Antifungal Properties of the Isolated Saponin Compounds from Different Plant Species

The crude extract from the stem bark of *Polyscias fulva* was fractionated and it resulted in the isolation of several known saponin compounds, in addition to one new saponin compound with chemical structure (3-O-[α-L-rhamnopyranosyl (1–2)-α-L-arabinopyranosyl]-28-O-[α-L-4-O-acetyl-rhamnopyranosyl (1–4)-β-D-glucopyranosyl-(1–6)-β-D-glucopyranosyl]-hederagenin) (Njateng et al. 2015). The isolated saponins were examined for their antimicrobial properties against different microbial species, including yeasts (*Candida albicans*, *C. krusei*, *C. parapsilosis*, *C. lucitaniae*, *C. glabrata*, *Cryptococcus neoformans*, and *C guilliermondii*) and dermatophytes (*Microsporum audouinii, Trichophyton rubrum, T. ajelloi, T. equinum, T. mentagrophytes, T. terrestre, T. violaceum, Microsporum gypseum, M. canis, M. ferrugeneum, Epidermophyton floccosum*. Among these compounds, 3-O-α-L-arabinopyranosyl-hederagenin and 3-O-[α-L-rhamnopyranosyl (1–2)-α-L-arabinopyranosyl]-hederagenin exhibited the highest antifungal activity against all tested pathogens with MIC values ranging from 0.78 to 100μg/ml (Njateng et al. 2015). Recently, a comparative study on the antifungal activities of saponin compounds against important crop pathogens using EC50 values demonstrated that aescin saponin exhibited the strongest antifungal activity against tested fungal pathogens. The effect of aescin saponin on plant–pathogen

Table 4.1 List of the saponin compounds isolated from different *Allium* species and their corresponding antifungal activity against different fungi pathogen and endophytes

Saponin compounds	Fungi used for antifungal assay	Reference
Aginoside	*Botrytis cinerea, B. squamosal, Fusarium oxysporum* f. sp. *cepa, F. oxysporum* f. sp. *lycopersici Colletotrichum gloeosporioides*	Mostafa et al. (2013)
Alliospiroside A Alliospiroside B	*Alternaria alternate, A. solani, A. tenuissima, B. cinerea, B. squamosa, C. acutatum, C. destructivum, C. graminicola, Curvularia lunata, Epicoccum nigrum, F. oxysporum* f. sp. *batatas, F. oxysporum* f. sp. *cepae, F. solani, F. proliferatum, F. verticillioides, Magnaporthe oryzae, Sclerotium cepivorum* and *Thanatephorus cucumeris*	Teshima et al. (2013)
Ceposide A Ceposide B Ceposide C	*Aspergillus niger, F. oxysporum* f. sp. *lycopersici, Trichoderma atroviride, A. alternate, B. cinerea, T. harzianum, Phomopsis* sp., and *Mucor* sp.	Lanzotti et al. (2012b)
Voghieroside A Voghieroside B Voghieroside C Voghieroside D Voghieroside E	*T. harzianum* and *B. cinerea*	Lanzotti et al. (2012a)
Alliospiroside A	*F. oxysporum* f. sp. *cepa*	Abdelrahman et al. (2017)
Minutoside A Minutoside B Minutoside C	*A. alternate, A. porri, B. cinerea, C. graminicola, F. solani, F. oxysporum* f. sp. *lycopersici Pythium ultimum, Rhizoctonia solani, Trichoderma harzianum*	Barile et al. (2007)
Proto-eruboside-B Eruboside-B	*Candida albicans*	Matsuura et al. (1988)
Persicosides A Persicosides B Persicosides C Persicosides D	*Penicillium italicum, A. niger, T. harzianum* and *B. cinerea*	Sadeghi et al. (2013)
β-chlorogenin	*C. albicans*	Mskhiladze et al. (2008)
β-chlorogenin	*F. culmorum*	Carotenuto et al. (1999)
Yayoisaponins A	*Mortierella ramanniana*	Sata et al. (1988)
Agigenin Ampeloside Bs1 Ampeloside Bf1	*A. niger* and *C. albicans*	Morita et al. (1988)

communication was evaluated through two different pathosystems, including *Arabidopsis thaliana* versus *Pseudomonas syringae* pv tomato (plant-bacterial interaction), and *Brassica napus* versus *Leptosphaeria maculans* (plant-fungi interaction) (Trdá et al. 2019). In addition, transcriptome analysis demonstrated that aescin induced *B. napus* defense through activation of the salicylic acid (SA) pathway and oxidative burst. Likewise, Aescin also inhibited the colonization of *A. thaliana* through the elicitation of SA-dependent immune mechanisms (Trdá et al. 2019). This aescin-induced defense mechanism enabled both *B. napus* and *A. thaliana* against *L. maculans*, and *P. syringae*, respectively, and the level of protection was comparable to the effect of fungicide application, providing the first clue regarding the ability of saponins to trigger plant immune responses (Trdá et al. 2019). In a similar study, the crude extract of dried pericarp of *Sapindus saponaria* L. fruits was subjected to column-chromatography, resulting in two pure triterpene acetylated saponins: 3-O-(4-acetyl-b-D-xylopyranosyl)-(1-3)-a-Lrhamnopyranosyl-(1-2)-a-L-arabinopyranosyl-hederagenin (1) and 3-O-(3,4-di-acetyl-b-D-xylopyranosyl)-(1-3)-a-L-rhamnopyranosyl-(1-2)-a-L-arabynopyranosyl-hederagenin (2). Then the isolated saponin compounds were further evaluated against different clinical pathogenic yeasts, revealing strong activity against *C. albicans*, *C. parapsilosis*, *C. glabrata*, and *C. tropicalis*, and among which *C. parapsilosis* was highly sensitive to saponin application compared with other saponins (Tsuzuki et al. 2007). Saponin rich-extracts derived from *Yucca schidigera*, *Balanites aegyptiaca* fruit, *Quillja saponaria* bark have been evaluated against several phytopathogenic fungi, including *F. oxysporum*, *Pythium ultimum*, *A. solani*, *Verticillium dahliae* and *Colletotrichum coccodes*. The antifungal effects of these saponin extracts using dose-dependent-fungi method was obtained. In general, crude saponins isolated from *B. aegyptiaca* fruit showed moderate (34.7%) to high (81.1%) level of antifungal activity against *A. solani and P. ultimum* respectively. However, growth inhibition was weak against *F. oxysporum*, *V. dahlia*, and *C. coccodes* (Chapagain et al. 2007). Similarly, *Quillja saponaria* bark saponin exhibited moderate growth inhibition (35.9–59.1%) against all tested fungi pathogens except *C. coccodes*, whereas *Yucca schidigera* saponin displayed highly significant (100%) to moderate (54.1%) growth inhibition on all the tested fungi. These results recommended that saponin can control outstandingly against these fungi (Chapagain et al. 2007). Another recent study demonstrated that methanol crude extract of *Trevesia palmata* showed potential antifungal properties against plant pathogenic fungi, such as *B. cinerea* and *Magnaporthe oryzae* (Kim et al. 2018). Based on antifungal activity fractions, there were five antifungal saponin compounds isolated from the methanol extract of *T. palmata*: including two new triterpene glycosides namely TPG1 and TPG5 (Kim et al. 2018). The chemical structure of TPG1 was hederagenin-3-O-β-D-glucopyranosyl-(1→3)-α-L-rhamnopyranosyl-(1→2)-α-L-rhamnopyranosyl-(1→2)-α-L-arabinopyranoside, and for TPG5 was 3-O-α-L-rhamnopyranosyl asiatic acid. In addition, to three known TPGs, including TPG2, TPG3 and TPG4 with chemical names macranthoside A, α-hederin, and TPG4 ilekudinoside D, respectively. An antifungal assay demonstrated that all the TGPs isolated except for TPG4 (ilekudinoside D), indicated the potential of having powerful antimycotic activities against the rice pathogen *M. oryzae* and tomato late blight, tomato grey mold, and wheat leaf rust, compared with the TGPs-non-treated control plants. The obtained results suggested that *T. palmata* can be a potential source for developing new natural fungicides.

References

Abdelrahman M, Hirata S, Ito SI, Yamauchi N, Shigyo M (2014) Compartmentation and localization of bioactive metabolites in different organs of *Allium roylei*. Biosci Biotechnol Biochem 78 (7):1112–1122

Abdelrahman M, El-Sayed M, Sato S, Hirakawa H, Ito S-I, Tanaka K, Mine Y, Sugiyama N, Suzuki Y, Yamauchi N, Shigyo M (2017) RNA-sequencing-based transcriptome and biochemical analyses of steroidal saponin pathway in a complete set of *Allium fistulosum-A. cepa* monosomic addition lines. PLoS One 12:e0181784

Barile E, Bonanomi G, Antignani V, Zolfaghari B, Sajjadi SE, Scala F et al (2007) Saponins from *Allium minutiflorum* with antifungal activity. Phytochemistry 68:596–603

Carotenuto A, Fattorusso E, Lanzotti V et al (1999) Spirostanol saponins of *Allium porrum* L. Phytochemistry 51:1077–1082

Chapagain B, Wiesman Z, Tsror L (2007) In vitro study of the antifungal activity of saponin-rich extracts against prevalent phytopathogenic fungi. Ind Crops Prod 26:109–115

da Cruz Cabral L, Fernandez Pinto V, Patriarca A (2013) Application of plant derived compounds to control fungal spoilage and mycotoxin production in foods. Int J Food Microbiol 166(1):1–14

Kim B, Han JW, Ngo MT, Dang QL, Kim JC, Kim H, Choi GJ (2018) Identification of novel compounds, oleanane- and ursane-type triterpene glycosides, from *Trevesia palmata*: their biocontrol activity against phytopathogenic fungi. Sci Rep 8:14522

Lanzotti V, Barile E, Antignani V, Bonanomi G, Scala F (2012a) Antifungal saponins from bulbs of garlic, *Allium sativum* L. var. Voghiera. Phytochemistry 78:126–134

Lanzotti V, Romano A, Lanzuise S, Bonanomi G, Scala F (2012b) Antifungal saponins from bulbs of white onion, *Allium cepa* L. Phytochemistry 74:133–139

Matsuura H, Ushiroguchi T, Itakura Y, Hayashi N, Fuwa T (1988) A furostanol glycoside from garlic, bulbs of *Allium sativum* L. Chem Pharm Bull 36:1347–5223

Matušinský P, Zouhar M, Pavela R, Nový P (2015) Antifungal effect of five essential oils against important pathogenic fungi of cereals. Ind Crops Prod 67:208–215

Morita T, Ushiroguchi T, Hayashi N et al (1988) Steroidal saponins from elephant garlic, bulbs of *Allium ampeloprasum*. Chem Pharm Bull. 36:3480–3486

Moses T, Papadopoulou KK, Osbourn A (2014) Metabolic and functional diversity of saponins, biosynthetic intermediates and semi-synthetic derivatives. Crit Rev Biochem Mol Biol 49:439–462

Mostafa A, Sudisha J, El-Sayed M, Ito SI, Ikeda T, Yamauchi N, Shigyo M (2013) Aginoside saponin, a potent antifungal compound, and secondary metabolite analyses from *Allium nigrum* L. Phytochem Lett 6:274–280

Mskhiladze L, Kutchukhidze J, Chincharadze D, Delmas F, Elias R, Favel A (2008) In vitro antifungal and antileishmanial activities of steroidal saponins from *Allium leucanthum* C. Koch—a Caucasian endemic species. Georgian Med News 154:39–43

Nicol RW, Traquair JA, Bernards MA (2002) Ginensosides as host resistance factors in American ginseng (*Panax quinquefolius*). Can J Bot 80:557–562

Njateng GSS, Du Z, Gatsing D, Donfack ARN, Talla MF, Wabo HK, Tane P, Mouokeu RS, Luo X, Kuiate JR (2015) Antifungal properties of a new terpernoid saponin and other compounds from the stem bark of *Polyscias fulva* Hiern (Araliaceae). BMC Complement Altern Med 15:25

Papadopoulou K, Melton RE, Leggett M, Daniels MJ, Osbourn AE (1999) Compromised disease resistance in saponin-deficient plants. Proc Natl Acad Sci U S A 96:12923–12928

Podolak I, Galanty A, Sobolewska D (2010) Saponins as cytotoxic agents: a review. Phytochem Rev 9(3):425–474

Porsche FM, Molitor D, Beyer M, Charton S, Andre C, Kollar A (2018) Antifungal activity of saponins from the fruit pericarp of *Sapindus mukorossi* against *Venturia inaequalis* and *Botrytis cinerea*. Plant Dis. 102(5):991–1000

Sadeghi M, Zolfaghari B, Senatore M, Lanzotti V (2013) Spirostane, furostane and cholestane saponins from Persian leek with antifungal activity. Food Chem 141:1512–1521

Saniewska A, Jarecka A, Bialy Z, Jurzysta M (2006) Antifungal activity of saponins originated from *Medicago hybrida* against some ornamental plant pathogens. Acta Agrobotanica 59 (2):51–58

Sata N, Matsunaga S, Fusetani N et al (1988) New antifungal and cytotoxic steroidal saponins from the bulbs of an elephant garlic mutant. Biosci Biotechnol Biochem 62:1904–1911

Schönbeck F, Schlösser E (1976) In: Heitefuss R, Williams PH (eds) Physiological plant pathology. Springer, Berlin, pp 653–678

Sreij R, Dargel C, Schweins R, Prevost S, Dattani R, Hellweg T (2019) Aescin-cholesterol complexes in DMPC model membranes: a DSC and temperature-dependent scattering study. Sci Rep 9(1):5542

Teshima Y, Ikeda T, Imada K, Sasaki K, El-Sayed MA, Shigyo M et al (2013) Identification and biological activity of antifungal saponins from shallot (*Allium cepa* L. Aggregatum group). J Agric Food Chem 61:7440–7445

Trdá L, Janda M, Macková D, Pospíchalová R, Dobrev PI, Burketová L, Matušinsky P (2019) Dual mode of the saponin aescin in plant protection: antifungal agent and plant defense elicitor. Front Plant Sci 10:1448

Tsuzuki JK, Svidzinski TIE, Shinobu CS, Silva LFA, Rodrigues-Filho E, Cortez DAG, Ferreira ICP (2007) Antifungal activity of the extracts and saponins from *Sapindus saponaria* L. Anais da Academia Brasileira de Ciências 79:1678–2690

VanEtten HD, Mansfield JW, Bailey JA, Farmer EE (1994) Plant Cell 6:1191–1192

Yang CR, Zhang Y, Jacob MR, Khan SI, Zhang YJ, Li XC (2006) Antifungal activity of C-27 steroidal saponins. Antimicrob Agents Chemother. 50:1710–1714

Yu Z-H, Ding X-Z, Xia L-Q et al (2013) Antimicrobial activity and mechanism of total saponins from *Allium chinense*. Food Sci. 34:75–80

Zimmer DE, Pedersen MW, McGuire DF (1967) A bioassay for alfalfa saponins using the fungus *Trichoderma viride*. Pers. ex. Fr. Crop Sci 7:223–224

Saponin-Detoxifying Enzymes

5

Abstract

Pathogenic fungi usually use different tactics to counteract induced and constitutive plant defense mechanisms that include degradation of any chemical compound and inhibition of plant triggered defenses by producing enzymes. Saponins as major bioactive compounds located in several monocot and dicot plant species and have been proposed to be involved in the defense of plants against pathogen outbreak. However, the capability of several pathogenic fungi to produce saponin-neutralizing enzymes would suggest that they play a major role in ascertaining the effect of interaction between plant and pathogen. Most of the saponin-detoxifying enzymes are glycosyl hydrolases, which catalyze hydrolysis of sugars from saponin aglycone that consists of a sugar chain attached to the C3 carbon, resulting in loss of saponin membranolytic properties and consequently loss of toxicity. In this chapter we will discuss and summarize different saponin-detoxifying enzymes and their effects in plant defense, as ultimate objective to increase crop plant productivity.

5.1 The Role of Saponin-Detoxifying Enzymes in Plant-Pathogen Interaction

Pathogenic fungi that cause diseases on host plants which contain saponin compounds are usually not susceptible to the toxicity of the saponin compounds compared with nonpathogens, indicating that resistance to saponin compounds is a prerequisite for pathogen infection (Arneson et al. 1968; Crombie et al. 1986; Suleman et al. 1996; Carter et al. 1999). However, the mechanisms of resistance against saponin compounds may vary depending on the type and composition of saponin compounds and pathogens. For example, Oomycete pathogens *Pythium* and *Phytophthora* are resistant to saponin compounds due to the lack of sterols in their membranes, which is essential for saponin attachments and activity (Arneson and

M. Abdelrahman, S. Jogaiah, *Bioactive Molecules in Plant Defense*,
https://doi.org/10.1007/978-3-030-61149-1_5

Durbin 1968a). The significance of the composition of sterol membrane was reported further by identifying the isolated sterol deficient mutants in *Neurospora crassa* and *Fusraium solani*, which showed greater resistance to the steroidal glycoalkaloid α-tomatine (Défago and Kern 1983; Senegupta et al. 1995). In addition, the *F. solani* sterol-deficient mutants caused infection in α-tomatine rich tomato fruits, whereas the wild-type *F. solani* shown pathogenicity to the tomato fruits with low α-tomatine, suggesting the importance of resistance by tomato pathogens towards α-tomatine (Défago and Kern 1983; Défago et al. 1983). In addition to the defect in sterol membrane composition, phytopathogenic fungi producing detoxifying enzymes to degrade the saponin compounds of their respective host plants, by removing the sugar molecules from the sugar chain attached to C-3 of the saponin aglycone (VanEtten et al. 1995; Osbourn 1996).

5.2 Detoxification of Tomato and Potato Saponins

Numerous solanaceous plant species produce glycosylated steroidal alkaloid and or saponin compounds (Roddick 1974). For example, in tomato plants, species like *Lycopersicon esculentum* are characterized by the presence of α-tomatine which is steroidal glycoalkaloid as the main saponin compound with potent antifungal activities against wide range of pathogens (Arneson and Durbin 1968a). However, *Septoria lycopersici* fungus, which causes tomato leaf spot disease, produces an extracellular enzyme namely, tomatinase that can hydrolyse glucose in the glycoside residues of α-tomatine to produce less inhibitory antifungal protein, namely β2-tomatine (Arneson and Durbin 1968b; Sandrock et al. 1995; Osbourn 1995; Sandrock and VanEtten 1998). On the other hand, a mutation in the tomatinase gene-encoding protein, resulted in loss of function and inability to convert α-tomatine to β2-tomatine, and thus increased pathogen sensitivity to α-tomatine antifungal activity (Martin-Hernandez et al. 2000; Sandrock and VanEtten 2001). In addition, *S. lycopersici* tomatinase-deficient mutants not only lack the ability to cause tomato leaf spot disease on tomato leaves, but also induced plant defense-related genes to express and triggered program cell death in the early infection stages (Martin-Hernandez et al. 2000). A similar finding was also reported *Nicotiana benthamiana* inoculated with three strains of *S. lycopersici* NEV, 16R and NY (Bouarab et al. 2002). Three strains of *S. lycopersici* pathogen caused disease lesions and tissue damage in the leaves of *N. benthamiana*, whereas *S. lycopersici* tomatinase-deficient mutants failed to cause any disease symptoms on *N. benthamiana* leaves (Bouarab et al. 2002). The above results suggested that tomatinase not only detoxify saponin compounds, but also used for suppressing plant defense mechanism to enable pathogen attacks (Bouarab et al. 2002). In general, tomatinase enzyme has been observed as detoxifying agent that enable fungal pathogen to grow in plants by degrading preformed host antibiotics (Morrissey and Osbourn 1999).

An investigation on the degradation of α-tomatine by *B. cinerea*, *S. lycopersici*, and *F. oxysporum* f. sp. *lycopersici* has been suggesting that it is common among the tomato pathogens that they are capable of hydrolyzing sugars from α-tomatine (Sandrock and VanEtten 1998). For example, the causal pathogen for *Alternaria* stem canker and *Corynespora* target spot disease were *Alternaria alternata* tomato pathotype and *Corynespora cassiicola*, respectively, were tested for sensitivity to α-tomatine saponin (Oka et al. 2006). The two strains of *A. alternata* and *C. cassiicola* pathogenic to tomato were not affected by α-tomatine treatments. On the other hand, α-tomatine treatment inhibited significantly the spore germination of *C. cassiicola* which is non-pathogenic to tomato (Oka et al. 2006). The α-tomatine resistance of both *A. alternata* and *C. cassiicola* was attributed to their ability to detoxify the α-tomatine into a less polar product, as an essential mechanism in order to colonize the host plant body, pathogens must produce host specific toxins (Oka et al. 2006). Potato (*Solanum tuberosum*) contains steroidal glycoalkaloids ~40–120 mg kg^{-1} of fresh weight (Friedman and Dao 1992). Among these steroidal glycoalkaloids, α-solanine and α-chaconine are the major two saponin compounds found in potato tuber (Friedman and McDonald 1997). The two steroidal saponin alkaloids have similar chemical structure, with 3-OH position of the steroidal glycoalkaloid solanidine is attached with trisaccharide. The antifungal activity of α-chaconine and α-solanine, against *Alternaria brassicicola*, *Phoma medicaginis*, *Ascobolus crenulatus*, and *Rhizoctonia solani* has been studied, and results indicated that, the two compounds produced synergistic antifungal effects against examined fungal pathogens. However, the magnitude of the antifungal activity varied depending on fungi species, saponin concentration and pH level (Fewell and Roddick 1993). In addition, three strains of filamentous fungi *Plectosphaerella cucumerina* have been isolated from potato sprouts, and these strains were able to hydrolyze α-chaconine into β_1-chaconine as first step in detoxification of filamentous fungi by removing the rhamnose (C1–C4) glucose linkage in α-chaconine to grow on potato sprouts (Oda et al. 2002). However, these strains were unable to detoxify α-solanine saponin. The partially purified enzyme was suggested to be a rhamnosidase specific for the hydrolysis of rhamnose in α-chaconine (Oda et al. 2002) (Fig. 5.1). In a recent study, α-chaconine was able to inhibit the growth of five fungal strains of *Alternaria alternata* AA001, *Pyrenophora teres* f. teres SK51, *Pyrenophora tritici-repentis* 331-2, *Mycosphaerella pinodes* Is.39 and *Mucor plumbeus* FUA5003 (Sánchez-Maldonado et al. 2016). However, there was resistance exhibited to α-chaconine by three fungal strains *A. niger* FUA5001, *P. roqueforti* FUA5005 and *F. graminearum* FG001 (Sánchez-Maldonado et al. 2016). In addition, it has been reported that the antifungal activity of α-solanine against ten strains from several species were different (Cipollini and Levey 1997). The ability of filamentous fungi to detoxify chaconine by removing sugars prevented the antifungal activity of the potato glycoalkaloids (Weltring et al. 1997; Oda et al. 2002). In addition to α-chaconine, an early study by Bushway et al. (1990) showed a high activity of detoxifying enzyme rhamnosidase isolated from the peel of potato cultivars 'Kennebec' and 'Wauseon' against α-solanine glycoalkaloids.

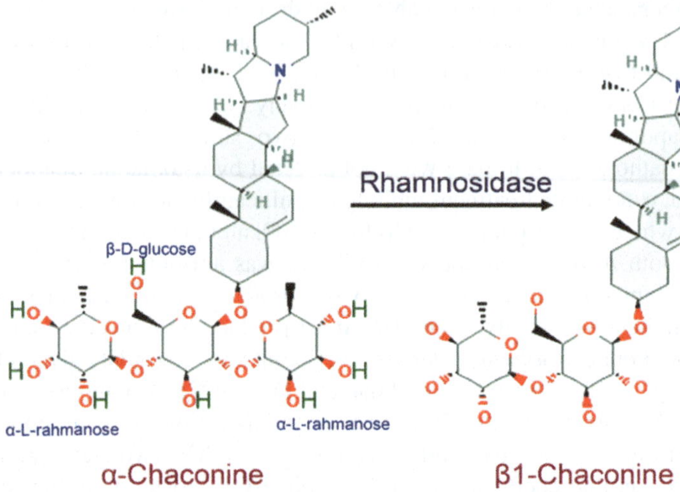

Fig. 5.1 Schematic figure of rhamnosidase detoxification activity against α-chaconine glycoalkaloids

5.3 Detoxification of Oat Saponins

Oat (*Avena sativa*) roots contain the triterpene saponin, namely avenacin that displayed antifungal properties against cereal root pathogens (Goodwin and Pollock 1954; Maizel et al. 1964). However, the oat root-infecting fungal pathogen *Gaeumannomyces graminis* var. *avenae*, was able to detoxify the triterpenoid avenacin saponins, by producing an extracellular enzyme avenacinase. By removing the D-glucose unit from the sugar chain attached to the saponin aglycone, the detoxifying enzyme avenacinase detoxifies avenacins. This detoxifying enzyme avenacinase was identified as a β-glucosyl hydrolase and is related to fungal cellobiose-degrading enzymes and xylosyl hydrolases (Bowyer et al. 1995; Margolles-Clark et al. 1996; Osbourn et al. 1995). In addition, the deficiency in the production of avenacinase protein in mutant fungi showed lack of ability to infect oat, indicating that the detoxification of avenacin saponin by avenacinase enzyme is an essential determinant of host range for *G. graminis* var. *avenae* (Bowyer et al. 1995). Avenacins have been reported as potent antifungal substance that protects oat roots against fungal attack. However, diploid oat species *A. longiglumis* that lack the production of avenacin saponins was susceptible to infection by *G. graminis* var. *tritici*, an avenacin sensitive fungi that usually unable to infect oats (Osbourn et al. 1994). In addition, the follow up study indicated that the susceptibility has been increased by the saponin-deficient (sad) mutants against to *G. graminis* var. *tritici* and other fungal pathogens (Papadopoulou et al. 1999).

Likewise, the oat foliar pathogen *Septoria avenae* showed resistance to the leaf oat saponin 26-desglucoavenacosides (26-DGAs) A and B (Osbourn et al. 1991;

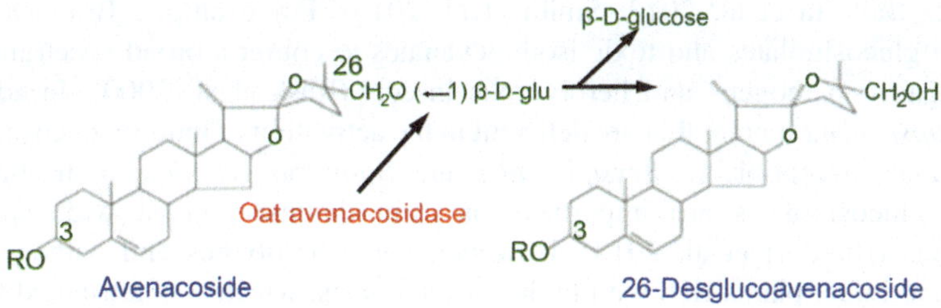

Fig. 5.2 Schematic representation of oat avenacosidase activity on avenacoside saponin to produce 26-Desglucoavenacoside

Osbourn et al. 1996; Wubben et al. 1996). In vitro studies suggested that *D. avenae* and *S. avenae* are able to detoxify oat saponin 26-DGAs by secreting an avenacosidase enzyme which in turn catalyzes the hydrolysis of both D-glucose and L-rhamnose molecules from the C-3 sugar chain in the 26-DGAs aglycon (Wubben et al. 1996) (Fig. 5.2). The detoxification of 26-DGAs seems to be a prerequisite for inducing the pathogenicity of *S. avenae* on oat leaves. On other hand, the wheat (*Triticum aestivum*)-attacking *S. avenae* isolates were unable to detoxify oat saponin 26-DGAs and subsequently could not infect oat leaf, suggesting that oat-attacking isolates are taxonomically different from wheat attacking isolates (Wubben et al. 1996).

5.4 Detoxification of Glucosinolates and Cyanogenic Glycosides

Glucosinolates are a large group of plant sulfur-containing compounds that are found in cruciferous vegetables, such as Brussels sprouts (*Brassica oleracea* var. *gemmifera*), broccoli (*B. oleracea* var. *italica*), and kale (*B. oleracea* var. *sabellica*) which have a distinctive pungent aroma and bitter taste (Chew 1988; Ciuffetti and VanEtten 1996; Fahey et al. 2001). Glucosinolate-containing vegetables provide health benefits that may extend well into the prevention of thoughtful diseases such as cancer (Bosetti et al. 2012). In addition to the Brassica genus and the cruciferous weed, glucosinolates have been identified in *Arabidopsis thaliana* (Duncan 1991). Chemically, thiohydroximate-O-sulfonate group is attached to glucose and indolyl, aralkyl or alkyl side chain of glucosinolates (Franco et al. 2016). Glucosinolates can be classified into three key classes based on the nature of the attached side chains, which may be derived from aliphatic, indolyl, or aralkyl a-amino acids. These compounds are usually stable in plant cell, however, when the glucosinolates found in plant cell is damaged, a β-thioglucosidase enzyme named myrosinase is released and can hydrolyze glucosinolates into β-d-glucose and an unstable aglycone; thiohydroximate-O-sulfonate (Halkier and Gershenzon 2006). Glucosinolates have been reported to be a significant component of plant defense system against pathogens and pests (Mithen 1992; Giamoustaris and Mithen 1995;

Martínez-Ballesta et al. 2013; Smith et al. 2016). For example, Brassica plants produce glucosinolates and toxic isothiocyanates to confer a broad spectrum resistance against pathogens and herbivorous insects (Rask et al. 2000). In addition, *Arabidopsis pen2* mutant that are deficient in the activation of indolic glucosinolates, were more susceptible to *Erysiphe pisi* and *Blumeria graminis*, indicating that indolic glucosinolates are important for plant defense against wide spectrum pathogens (Buxdorf et al. 2013). However, some herbivores and pathogens can digest and infect plants rich with high glucosinolates, which was attributed to their ability to metabolism of glucosinolates into non-toxic forms. For instance, specialized herbivores use glutathione-dependent mercapturic acid pathway to conjugate glucosinolates hydrolysed products and thereby, deactivating those (Jeschke et al. 2017). Buxdorf et al. (2013) reported that the hydrolyzed products of indolic glucosinolates can differentiate between plant responses to *B. cinerea* and plant responses to *A. brassicicola*, while *B. cinerea* was sensitive to indolic glucosinolates, *A. brassicicola* showed high tolerability to the presence of glucosinolates and their breakdown products. In addition, fungal pathogen such as *Sclerotinia sclerotiorum* is able to infect many plants rich with glucosinolate, by activating the glucosinolate-myrosinase system to produce isothiocyanates, then *S. sclerotiorum* metabolizes isothiocyanates via conjugating isothiocyanates to glutathione, a non-toxic form; or through hydrolysis of isothiocyanates into amines using isothiocyanate hydrolase enzyme (Chen et al. 2020). The isothiocyanate hydrolase enzyme in the presence of the toxins enhances fungal growth, and adds to the virulence of *S. sclerotiorum* on plants containing glucosinolate (Chen et al. 2020).

There are similarities between glucosinolates and cyanogenic glycosides, as both metabolites are synthesized from amino acids via oxime intermediates but homologous enzymes may control some of the biosynthetic steps involved as proposed (Poulton and Moller 1993). Several important crops contain cyanogenic glycosides, including sorghum (*Sorghum bicolor*), cassava (*Manihot esculenta*), bamboo (*Bambusa vulgaris*), cocoyam (*Colocasia esculenta* L. and *Xanthosoma sagittifolium* L.) andapple (*Malus domestica*). Cyanogenic glycosides synthesized by the conversion of amino acids to oximes, and the latter glycosylated into cyanogenic glycosides (Poulton 1988; Davis 1991). Cyanogenic glycosides and their derivatives have amino acid-derived aglycones, which naturally degrade to release highly toxic hydrogen cyanide. Currently, cyanogenic glycosides have been identified in more than 200 plant species including monocots, dicots, ferns, and gymnosperms (Poulton 1988; Davis 1991; Hurst et al. 2008). The accumulation of hydrogen cyanide, a potent respiratory toxin, represents chemical barrier that shields cyanogenic plants against herbivores and pathogens. However, pathogenic fungi that can infect cyanogenic plants are usually able to tolerate hydrogen by inducing enzymatic detoxification systems (Fry and Myers 1981). The association between the capacity to produce the cyanide-detoxifying enzyme, cyanide hydralase and the capability to infect cyanogenic plants has been tested for fungal pathogen, *Gloeosporioides sorghi* on the cyanogenic plant sorghum (Yue et al. 1998; Morrissey and Osbourn 1999). For instance, under in vitro conditions, cyanide hydralase-deficient mutants of *G. sorghi* generated by disruption of targeted gene were more sensitive than wild-type fungi to hydrogen cyanide, supporting that the enzyme could confer resistance to hydrogen cyanide (Morrissey and Osbourn 1999). However, pathogenicity to sorghum was unaffected, indicating either that the fungus has more tolerance towards cyanide which supports its growth in sorghum tissue by alternative means or the capability to tolerate hydrogen cyanide is not required for infection of sorghum by *G. sorghi* (Morrissey and Osbourn 1999). The cyanogenic glucoside levels significantly reduced to 62.7% (SD 2.8) of the initial value as a result of incubating disinfected cassava root pieces. The incubation of cassava root with different fungi, including *Geotrichum candidum*, *Mucor racemosus*, *Neurospora sitophila*, *Rhizopus oryzae* and *Rhizopus stolonifer*, or a *Bacillus* sp. resulted in substantial drop to 29.8% in the cyanogenic glucoside levels, compared to the level obtained under non-inoculated incubation (Essers et al. 1995). Among the tested strains, *N. sitophila* reduced cyanogenic glucoside levels most effectively, followed by *R. stolonifer* and *R. oryzae*, which was attributed to their ability to produce cyanogenic hydrolase enzyme (Essers et al. 1995).

5.5 Detoxification of *Allium* Saponins

With respect to *Allium* saponins, the early studies indicated that *Botrytis cinerea* and *Trichoderma atroviride* were more sensitive to three saponin compounds isolated from onion, namely (25R)-furost-5(6)-en-1β,3β,22α,26-tetraol 1-O-β-D-xylopyranosyl 26-O-α-D-rhamnoyranosyl-(1→2)-O-β-D-galactopyranoside (ceposide A), (25R)-furost-5(6)-en-1β,3β,22α,26-tetraol 1-O-β-D-xylopyranosyl 26-O-α-D-rhamnoyranosyl-(1→2)-O-β-D-glucopyranoside (ceposide B), and (25R)-furost-5(6)-en-1β,3β,22α,26-tetraol 1-O-β-D-galactopyranosyl 26-O-α-D-rhamnoyranosyl-(1→2)-O-β-D-galactopyranoside (ceposide C) (Lanzotti et al. 2011). In contrast, *Fusarium oxysporum* f. sp. *lycopersici*, *Sclerotium cepivorum* and *Rhizoctonia solani* were affected insignificantly by saponins, which were attributed to their ability to detoxify these saponin compounds through the enzymatic hydrolysis of the sugar chains attached to the saponin aglycon (Lanzotti et al. 2011; Lanzotti 2012). Likewise, Alliospiroside A, a saponin exbitied potent antifungal activity against different phytopathogens, whereas many of the *Fusarium* pathogens were more tolerant to the saponin compounds (Teshima et al. 2013). Saponin compounds namely minutoside A, minutoside B and minutoside C along with two sapogenins, alliogenin and neoagigenin were isolated and identified from *A. minutiflorum*, a bulbous perennial plant known as wild onion that is used in Persia for food preparation (Lanzotti 2012). Minutoside A, B, and C, and the two sapogenin neoagigenin and alliogenin were examined for their broad-spectrum antimicrobial activity against several fungal and bacterial microorganisms. All saponin compounds exhibited a significant antifungal activity based on their concentration and minutoside B showed the highest activity followed by minutoside C, neoagigenin, alliogenin and minutoside A (Lanzotti 2012). However, none-of the examined saponin or sapogenin compounds showed any antibacterial activity.

Authors suggested that structure-activity relationships might be attributed to the observed antifungal properties (Lanzotti 2012). For example, spirostanol saponin minutoside B was more powerful in terms of bioactivity compared with furostanol saponin minutoside A, suggesting that spirostanol-type aglycone is much important for the antifungal activity. In addition, the two *Trichoderma harzianum* strains were more sensitive than the examined pathogenic fungi, which was evident by complete inhibition of the *Trichoderma harzianum* strains at the lowest concentration by minutosides B and C, and neoagigenin (Lanzotti 2012). In contrast, the *Pythium ultimum* Oomycetes showed more resistance to all investigated saponin compounds (Lanzotti 2012). The ability of *P. ultimum* tolerance was interlinked with the sterols lacking in the membrane of Oomycetes, which is essential for saponins to express their antifungal activity by forming complex with sterols, inflict damage in the fungi membrane (Morrissey and Osbourn 1999). The potent antifungal activity of *A. minutiflorum* saponins, especially those demonstrated by minutoside B, suggests that these saponin compounds, alone or in combination, may act as chemical barriers to fungal attacks. However, many fungi may attack plants by producing saponin-detoxifying enzymes that degrade saponins into non-toxic compounds (Sandrock and VanEtten 1998). For example, *Armillaria mellea*, a plant fungal pathogen also showed resistance mechanism that is able to degrade the antifungal isoflavone genistein into non-toxic metabolites (Curir et al. 2006). Therefore, the high sensitivity of the two *T. harzianum* strains towards *A. minutiflorum* saponins could be associated with their lower ability to detoxify these saponin compounds.

Saponin-enzyme relationships have been also investigated not only from pathogen detoxifying enzymes, but through the inhibitory effects of saponins against several functional enzymes. For example, saponin fraction obtained from the methanol extract of *A.chinense* was able to inhibit cAMP phosphodiesterase (cAMP PDE) enzyme that breaks a phosphodiester bond, and sodium–potassium adenosine triphosphatase (Na^+/K^+ATPase) enzyme responsible for establishing Na^+ and K^+ concentration gradients across the plasma membrane at low concentration (Kuroda et al. 1995). Likewise, (25R,S)-5α-spirostane-3β-ol tetrasaccharide saponin was able to inhibit the two enzymes cAMP PDE and Na^+/K^+ATPase (Kuroda et al. 1995). Similarly, saponins isolated from *A. giganteum* was able to inhibit cAMP PDE (Mimaki et al. 1994), whereas saponins isolated from *A. cepa* and *A. karataviense* were able to inhibit the activity of highly purified porcine kidney Na^+/K^+ATPase enzymein the concentration range from 1×10^{-4} to 1×10^{-7} M (Mirsalikhova et al. 1993). Moreover, it was revealed that alliospirosides A and B were both uncompetitive enzyme inhibitors, while alliospiroside D showed competitive enzyme inhibitor.

5.6 Conclusion

Several pathogenic fungi showed considerable resistance to saponin compounds, however the detailed mechanisms of this resistance is still not clear. Although the ability of these pathogenic fungi in the production of detoxifying-enzyme is one of the key protective for such bioactivity, isolation and identification if the saponin-detoxifying enzymes is high priority to better understand the plant-pathogen relationship. Specifically, *Fusarium* pathogens showed higher resistance to saponins isolated from different *Allium* species, but so far, the detoxifying enzymes produced by fusarium pathogens has not been identified and remain a future task.

References

Arneson PA, Durbin RD (1968a) The sensitivity of fungi to α-tomatine. Phytopathology 58:536–537

Arneson PA, Durbin RD (1968b) Studies on the mode of action of tomatine as a fungitoxic agent. Plant Physiol 43:683–686

Bosetti C, Filomeno M, Riso P et al (2012) Cruciferous vegetables and cancer risk in a network of case-control studies. Ann Oncol 23:2198–2203

Bouarab K, Melton R, Peart J, Baulcombe D, Osbourn A (2002) A saponin-detoxifying enzyme mediates suppression of plant defences. Nature 418:889–892

Bowyer P, Clarke BR, Lunness P, Daniels MJ, Osbourn AE (1995) Host range of a plant pathogenic fungus determined by a saponin detoxifying enzyme. Science 267:371–374

Bushway AA, Bushway RJ, Kim CH (1990) Isolation, partial purification, and characterization of a potato peel α-solanine cleaving glycosidase. Am Potato J 67:233–238

Buxdorf K, Yaffe H, Barda O, Levy M (2013) The effects of glucosinolates and their breakdown products on necrotrophic fungi. PLoS One 8:e70771

Carter JP, Spink J, Cannon PF, Daniels MJ, Osbourn AE (1999) Isolation, characterization, and avenacin sensitivity of a diverse collection of cereal-root-colonizing fungi. Appl Environ Microbiol 65:3364–3372

Chen J, Ullah C, Reichelt M, Beran F, Yang Z-L, Gershenzon J, Hammerbacher A, Vassão DG (2020) The phytopathogenic fungus *Sclerotinia sclerotiorum* detoxifies plant glucosinolate hydrolysis products via an isothiocyanate hydrolase. Nat Comm 11:3090

Chew FS (1988) Biological effects of glucosinolates. Am Chem Soc Symp Ser 380:155–181

Cipollini ML, Levey DJ (1997) Why are some fruits toxic? Glycoalkaloids in solanum and fruit choice by vertebrates. Ecology 78:782–798

Ciuffetti LM, VanEtten HD (1996) Virulence of a pisatin demethylase-deficient *Nectria haematococca* MPVI isolate is increased by transformation with a pisatin demethylase gene. Mol Plant Microbe Interact 9:787–792

Crombie WML, Crombie L, Green JB, Lucas JA (1986) Pathogenicity of the take-all fungus to oats: its relationship to the concentration and detoxification of the four avenacins. Phytochemistry 25:2075–2083

Curir P, Dolci M, Corea G, Galeotti F, Lanzotti V (2006) The plant antifungal isoflavone genistein is metabolized by *Armillaria mellea* Vahl to give non-fungitoxic products. Plant Biosyst 140:156–162

Davis RH (1991) Glucosinolates. In: D'Mello JP, Duffus CM, Duffus JH (eds) Toxic substances in crop plants. The Royal Society of Chemistry, Cambridge, UK, pp 202–225

Défago G, Kern H (1983) Induction of *Fusarium solani* mutants insensitive to tomatine, their pathogenicity and aggressiveness to tomato fruits and pea plants. Physiol Plant Pathol 22:29–37

Défago G, Kern H, Sedlar L (1983) Genetic analysis of tomatine insensitivity, sterol content and pathogenicity for green tomato fruits in mutants of *Fusarium solani*. Physiol Mol Plant Pathol 22:39–43

Duncan AJ (1991) Glucosinolates. In: D'Mello JP, Duffus CM, Duffus JH (eds) Toxic substances in crop plants. The Royal Society of Chemistry, Cambridge, UK, pp 126–147

Essers AJA, Jurgens CMGA, Nout MJR (1995) Contribution of selected fungi to the reduction of cyanogen levels during solid substrate fermentation of cassava. Int J Food Microbiol 26:251–257

Fahey JW, Zalcmann AT, Talalay P (2001) The chemical diversity and distribution of glucosinolates and isothiocyanates among plants. Phytochemistry 56:5–51

Fewell AM, Roddick JG (1993) Interactive antifungal activity of the glycoalkaloids α-solanine and α-chaconine. Phytochemistry 33:323–328

Franco P, Spinozzi S, Pagnotta E, Lazzeri L, Ugolini L, Camborata C et al (2016) Development of a liquid chromatography—electrospray ionization—tandem mass spectrometry method for the simultaneous analysis of intact glucosinolates and isothiocyanates in Brassicaceae seeds and functional foods. J Chromatogr A 1428:154–161

Friedman M, Dao L (1992) Distribution of glycoalkaloids in potato plants and commercial potato products. J Agric Food Chem 40:419–423

Friedman M, McDonald GM (1997) Potato glycoalkaloids: chemistry, analysis, safety, and plant physiology. Crit Rev Plant Sci 16:55–l 32

Fry WE, Myers DF (1981) Hydrogen cyanide metabolism by fungal pathogens of cyanogenic plants. In: Vennesland B, Knowles CJ, Conn EE, Westley J, Wissing F (eds) Cyanide in biology. Academic, London, pp 321–334

Giamoustaris A, Mithen R (1995) The effect of modifying the glucosinolate content of leaves of oilseed rape (*Brassica napus ssp. oleifera*) on its interaction with specialist and generalist pests. Ann Appl Biol 126:347–363

Goodwin RH, Pollock BM (1954) Studies on roots. I. Properties and distribution of fluorescent constituents in Avena roots. Am J Bot 4:516–520

Halkier BA, Gershenzon J (2006) Biology and biochemistry of glucosinolates. Ann Rev Plant Biol 57:303–333

Hurst WJ, Glinski JA, Miller KB, Apgar J, Davey MH, Stuart DA (2008) Survey of the trans-resveratrol and trans-piceid content of cocoa-containing and chocolate products. J Agric Food Chem 56:8374–8378

Jeschke V et al (2017) How glucosinolates affect generalist Lepidopteran larvae: growth, development and glucosinolate metabolism. Front Plant Sci 8:1995

Kuroda M, Mimaki Y, Kameyama A et al (1995) Steroidal saponins from Allium chinense and their inhibitory activities on cyclic AMP phosphodiesterase and $Na^{+}K^{+}$ ATPase. Phytochemistry 40:1071–1076

Lanzotti V (2012) Bioactive polar natural compounds from garlic and onions. Phytochem Rev 11:179–196

Lanzotti V, Romano A, Lanzuise S, Bonanomi G, Scala F (2011) Antifungal saponins from bulbs of white onion, *Allium cepa* L. Phytochemistry 74:133–139

Maizel JV, Burkhardt HJ, Mitchell HK (1964) Avenacin, an antimicrobial substance isolated from *Avena sativa*. I. Isolation and antimicrobial activity. Biochemistry 3:424–431

Margolles-Clark E, Tenkanen M, Nakari-Setälä T, Penttila M (1996) Cloning of genes encoding β-l-arabinofuranoside and β-xylosidase from *Trichoderma reesei* by expression in *Saccharomyces cerevisiae*. Appl Environ Microbiol 62:3840–3846

Martínez-Ballesta MC, Moreno DA, Carvajall M (2013) The physiological importance of glucosinolates on plant response to abiotic stress in Brassica. Int J Mol Sci 14:11607–11625

Martin-Hernandez AM, Dufresne M, Hugouvieux V, Melton R, Osbourn A (2000) Effects of targeted replacement of the tomatinase gene on the interaction of *Septoria lycopersici* with tomato plants. Mol Plant Microbe Interact 13:1301–1311

Mimaki Y, Nikaido T, Matsumoto K et al (1994) New steroidal saponins from the bulbs of *Allium giganteum* exhibiting potent inhibition of cAMP phosphodiesterase activity. Chem Pharm Bull 42:710–714

Mirsalikhova NM, Kravets SS, Sokolova SF et al (1993) Inhibition of highly purified porcine kidney Na, K-ATPase by steroid glycosides of the spirostan and furostan series and a study of structure–activity relationships. Chem Nat Comp 29:490–497

Mithen R (1992) Leaf glucosinolate profiles and their relationship to pest and disease resistance in oilseed rape. Euphytica 63:71–83

Morrissey JP, Osbourn AE (1999) Fungal resistance to plant antibiotics as a mechanism of pathogenesis. Microbiol Mol Biol Rev 63:708–724

Oda Y, Saito K, Ohara-Takada A, Mori M (2002) Hydrolysis of the potato glycoalkaloid α-chaconine by filamentous fungi. J Biosci Bioeng 94:321–325

Oka K, Okubo A, Kodama M, Otani H (2006) Detoxification of α-tomatine by tomato pathogens *Alternaria alternata* tomato pathotype and *Corynespora cassiicola* and its role in infection. J Gen Plant Pathol 72:152–158

Osbourn AE (1996) Saponins and plant defence-a soap story. Trends Plant Sci 1:4–9

Osbourn AE, Clarke BR, Dow JM, Daniels MJ (1991) Partial characterization of avenacinase from *Gaeumannomyces graminis* var. avenae. Physiol Mol Plant Pathol 38:301–312

Osbourn AE, Clarke BR, Lunness P, Scott PR, Daniels MJ (1994) An oat species lacking avenacin is susceptible to infection by *Gaeumannomyces graminis* var. *tritici*. Physiol Mol Plant Pathol 45:457–467

Osbourn AE, Bowyer P, Lunness P, Clarke B, Daniels M (1995) Fungal pathogens of oat roots and tomato leaves employ closely related enzymes to detoxify different host plant saponins. Mol Plant Microbe Interact 8:971–978

Osbourn AE, Bowyer P, Daniels MJ (1996) Saponin detoxification by plant pathogenic fungi. In: Waller GR, Yamasaki K (eds) Saponins used in traditional and modern medicine. Advances in experimental medicine and biology, vol 404. Springer, Boston, MA

Papadopoulou K, Melton RE, Leggett M, Daniels MJ, Osbourn AE (1999) Compromised disease resistance in saponin-deficient plants. PNAS 96:12923–12928

Poulton JE (1988) Localization and catabolism of cyanogenic glycosides. Ciba Found Symp 140:67–91

Poulton JE, Moller BL (1993) Glucosinolates. Methods Plant Biochem 9:209–237

Rask L et al (2000) Myrosinase: gene family evolution and herbivore defense in Brassicaceae. Plant Mol Biol 42:93–113

Roddick J (1974) The steroidal glycoalkaloid tomatine. Phytochemistry 13:9–25

Sánchez-Maldonado AF, Schieber A, Gänzle MG (2016) Antifungal activity of secondary plant metabolites from potatoes (*Solanum tuberosum* L.): glycoalkaloids and phenolic acids show synergistic effects. J Appl Microbiol 120:955–965

Sandrock RW, VanEtten HD (1998) Fungal sensitivity to and enzymatic degradation of the phytoanticipin α-tomatine. Phytopathology 88:137–143

Sandrock RW, VanEtten HD (2001) The relevance of tomatinase activity in pathogens of tomato: disruption of the β2-tomatinase gene in *Colletotrichum coccodes* and *Septoria lycopersici* and heterologous expression of the *Septoria lycopersici* β2-tomatinase in *Nectria haematococca*, a pathogen of tomato fruit. Physiol Mol Plant Pathol 58:159–171

Sandrock RW, DellaPenna D, VanEtten HD (1995) Purification and characterization of β2-tomatinase, an enzyme involved in the degradation of α tomatine and isolation of the gene encoding β2-tomatinase from *Septoria lycopersici*. Mol Plant Microbe Interact 8:960–970

Senegupta S, Prasanna TB, Kasbekar DP (1995) Sterol 14,15 reductase (erg-3) mutations switch the phenotype of *Neurospora crassa* from sensitivity to the tomato saponin α-tomatine to sensitivity to the pea phytoalexin pisatin. Fungal Genet Newsl 42:71–72

Smith JD, Woldemariam MG, Mescher MC, Jander G, De Moraes CM (2016) Glucosinolates from host plants influence growth of the parasitic plant *Cuscuta gronovii* and its susceptibility to aphid feeding. Plant Physiol 172:181–197

Suleman P, Tohamy AM, Saleh AA, Madkour MA, Straney DC (1996) Variation in sensitivity to tomatine and rishitin among isolates of *Fusarium oxysporum* f.sp. *lycopersici*, and strains not pathogènic to tomato. Physiol Mol Plant Pathol 48:131–144

Teshima Y et al (2013) Identification and biological activity of antifungal saponins from shallot (*Allium cepa* L. Aggregatum Group). Agric Food Chem (31):7440–7445

VanEtten HD, Sandrock RW, Wasmann CC, Soby SD, McCluskey K, Wang P (1995) Detoxification of phytoanticipins and phytoalexins by phytopathogenic fungi. Can J Bot 73:S518–S525

Weltring KM, Wessels J, Geyert R (1997) Metabolism of the potato saponins α-chaconine and α-solanine by *Gibberella pilicaris*. Phytochemistry 46:1005–1009

Wubben JP, Price KR, Daniels MJ, Osbourn AE (1996) Detoxification of oat leaf saponins by *Septoria avenae*. Phytopathology 86:986–992

Yue Q, Bacon CW, Richardson MD (1998) Biotransformation of 2-benzoxazolinone and 6-methoxy-benzoxazolinone by *Fusarium moliliforme*. Phytochemistry 48:451–454

Isolation and Characterization of Triterpenoid and Steroidal Saponins

6

Abstract

Saponins are broadly dispersed natural products in the plant kingdom with massive structural and functional diversity, and therefore being regarded as an active components in medicinal plants. Saponin compounds have a significant roles in pharmaceutical industry, however the specific roles of saponins in plant defense as well as other biological process are remain underexplored. Saponins are glycosides of steroids, triterpenes or alkaloids, which are primarily found in roots and shoots of different plant species. Therefore, saponin compounds can be classified into steroidal, triterpenoidal or alkaloidal saponin depending on the nature of their aglycone structure. In this chapter, we will discussed the two major saponin classes, including triterpene saponins and steroidal saponins and their biological activities in pharmaceutical industries and plant-microbe interactions. In addition, saponin biosynthesis pathway and methods of induction of saponin contents will be also covered in this chapter.

6.1 Chemistry of Saponins

Several saponin compounds are found in different plant species in the form of glycosides of complex alicyclic compounds. However, some plants species might have very little amount or doesn't produce saponins, while in other plant species steroidal saponins or triterpene saponins are predominant (Mostafa et al. 2013; Abdelrahman et al. 2014, 2017b; Fanani et al. 2019). Acid hydrolysis of saponin compound produces two major parts, sugar moiety and aglycone (Abdelrahman et al. 2017a). Based on the structure of aglycone/sapogenin, saponins can be divided into three major groups, including (1) triterpenoid glycosides, (2) steroid glycosides and (3) alkaloid glycosides (Fig. 6.1) (Abed El Aziz et al. 2019). Based on their chemical structure, saponin compounds can possess either a triterpenoid (C30) or steroidal (C27) aglycone skeleton, with different numbers of sugar chains attached at

M. Abdelrahman, S. Jogaiah, *Bioactive Molecules in Plant Defense*,
https://doi.org/10.1007/978-3-030-61149-1_6

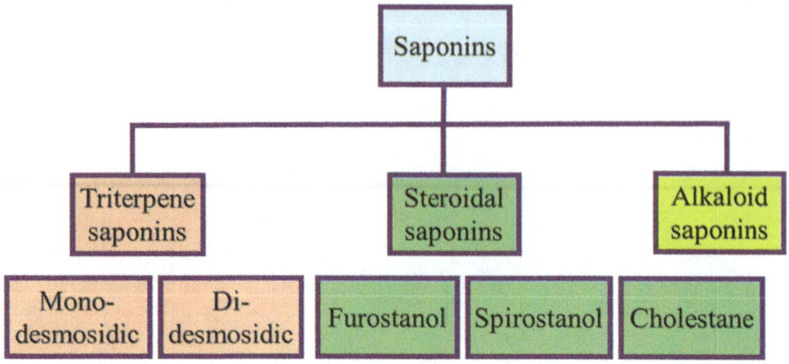

Fig. 6.1 Schematic classification of saponin diversification in different plant kingdom

various positions. For instance, triterpene saponins are mainly distributed in dicotyledonous angiosperms, and consisted of three monoterpenes with C30 carbon atoms allocated on six isoprene units. Triterpene saponins can be also subdivided into monodesmosidic or didesmosidic types (Fig. 6.1), based on the number of sugar moieties attached to the core aglycone (Fanani et al. 2019). For example, bidesmosidic triterpene glycosides have two sugar chains, one sugar chain attached at C3 position of the aglycone, and the other sugar chain either attached through an ether linkage at C24 position or ester linkage at C28, whereas monodesmosidic triterpene glycosides have a single sugar chain, attached at C-3 position (Fanani et al. 2019). Due to structural diversity of triterpene saponins, these compounds are considered an important bio-resources for novel drug discoveries (Geisler et al. 2013; Vo et al. 2017). However, due to the limited information regarding the molecular mechanisms underlying triterpenoid structural diversity, the potential of triterpene saponin engineering and application have been hampered. The second major class in saponin compounds is steroid glycosides, which are mainly found in monocotyledonous angiosperms, specifically in the onion (Amaryllidaceae), asparagus (Asparagaceae), yam (Dioscoreaceae), solanum (Solanaceae) and lily (Liliaceae) families (Challinor and De Voss 2013). In general, steroidal saponins are generally classified into two sub-classes, namely furostanol and spirostanol saponins, in addition to a third structural sub-class namely cholestane, which is presumed to be the early precursor of all steroidal saponins (Challinor and De Voss 2013). Spirostanol aponins are characterized by a bicyclic spiroacetal moiety at C22 position, which contains both the steroid E and F rings (Challinor and De Voss 2013). Spirostanol saponins are classically monodesmosidic in nature, thus, they are having one sugar units attached into the sapirostanol aglycone at one position (Abugabr Elhag et al. 2018). For example, dioscin, an archetypal spirostanol saponin consisted of the spirostanol aglycone diosgenin with a branched trisaccharide chain at C3 position. (Vincken et al. 2007). On the other hand, a hemiacetal, methyl acetal, or $\Delta^{20(22)}$ unsaturation are characteristic features for the structure of furostanol saponins (Sparg et al. 2004). For example, protodioscin a classic example of furostanol saponin type possess an O-linked β-d-glucose residue linked at C26 position, which prevents the cyclization and subsequently the formation of the steroidal F ring as in spirostanol saponins (Sparg et al. 2004). Although aglycon structure is very important for saponin diversification, especially the stereochemistry and oxygenation platforms, the main source of diversity in furostanol and spirostanol saponin structure is also attributed to the variation in the total number of the linked saccharide moieties (Challinor and De Voss 2013). For instance, both protogracillin and gracillin have similar structure with protodioscin and dioscin in term of aglycone, respectively, however, they differ in trisaccharide moiety linked at C3 position (Kräutler et al. 2008; Challinor et al. 2012). In addition, to furostanol and spirostanol saponins, a third the open-chain steroidal glycosides with skeleton structure closely resemble to cholesterol can be also identified. Openchain steroidal glycosides consisted of a C27 cholesterol-derived steroidal skeleton attached with various numbers of sugar residues at various positions. In addition, Openchain steroidal glycosides lack the heterocyclic ring(s) derived from the C17 sidechain which is characterisitic features in both spirostanol and furostanol saponins (Challinor and De Voss 2013). Therefore, the open-chain saponin glycosides considered as a distinct class of steroidal saponins (Challinor and De Voss 2013). A total of 150 open-chain steroidal saponin glycosides have been isolated and identified from different plant species. The main source of structural diversity of this class is mostly generated from the differences in the arrangement of oxygenation of the aglycone and the number of the attached monosaccharide units (Challinor and De Voss 2013). In general, most of the plant derived steroidal saponins with openchain steroidal glycosides possessed the C3 oxygenation that is found in their early biosynthetic ancestor cholesterol through the cyclization of 2,3-oxidosqualene (Challinor and De Voss 2013).

6.2 Triterpene Saponins

6.2.1 Triterpene Saponins in Leguminous Plants

Legumes are very economically important crops characterized by high protein contents and massive array of natural products, including anthocyanins, lignin, isoflavonoids and saponins (Lei et al. 2019). Triterpene saponins isolated from leguminous plant species are characterized by a triterpene aglycone attached with one, two, or sometimes three saccharide chains with different size and complexity. LC-MS-based metabolomics is well suited for the analysis of saponin compounds in many legume species including alfalfa (*Medicago sativa*), clover (*Trifolium hybridum),* soybean, *M. truncatula*, and *M. arborea* (Bialy et al. 1999; Huhman et al. 2005; Tava et al. 2005; Kapusta et al. 2005a, b; Pollier et al. 2011; Perez et al. 2013). For example, saponins in 12 different *Medicago* species were analyzed, and the levels of saponin contents were ranged from 0.38 to 1.35% per dry weight, according to the species (Tava and Pecetti 2012). Moreover, variations in the aglycone residues were observed among the investigated 12 *Medicago* species, for instance, some saponin compounds such as hederagenin and bayogenin were detected in the all examined 12 *Medicago* species, while other saponin aglycons such as zanhic acid and medicagenic acid were species-specific (Tava and Pecetti 2012). Likewise, five azukisapogenol saponin glycosides have been isolated from the tissues of clover plants, and their chemical structures have been clarified using 1D and 2D NMR spectrometric combined with HRESIMS and ESI-MS/MS (Pérez et al. 2013). Azukisapogenol glycosides 3-O-[-α-L-arabinopyranosyl(1→2)]-β-D-glucuronopyranosyl azukisapogenol (referred as compound 1), 3-O-[-β-D-glucuronopyranosyl(1→2)-β-D-glucuronopyranosyl]-29-O-β-D-glucopyranosyl azukisapogenol (compound 2), and 3-O-[-α-L-arabinopyranosyl(1→2)-β-D-glucuronopyranosyl]-29-O-β-D-glucopyranosyl azukisapogenol (compound 3) were identified as new compounds in clover plant, whereas the two other compounds are known compounds (Pérez et al. 2013). The sugar moieties in the all isolated compounds of the β-d-glucuronic acid possessed a monosaccharide unit attached at C3 position of the aglycone backbone (Pérez et al. 2013). This sugar structure feature us similar to the saponin profiles previously reported from other *Trifolium* species, which can be recognized as a chemotaxonomic character in the *Trifolium* genus (Pérez et al. 2013). However, a comprehensive phytochemical analysis in different organs of the plant has to be conducted to confirm this hypothesis. Bidesmosidic saponins are recognized by the presence of a β-d-glucose residue attached at the C29 position, as a glycosidic esterification. Similarly, azukisapogenol a triterpenoid aglycone, has been reported for the first time in this *Trifolium* genus. In a recent study, a total of 201 *Medicago truncatula* ecotypes originated from 14 different Mediterranean countries were analyzed for their saponin profiles using UHPLC-MS to deliver information for a genome-wide association and facilitate the germplasm selection for saponin biosynthesis (Lei et al. 2019). Saponin contents were significantly different among the investigated *M. truncatula* ecotypes. For instance, European *M. truncatula* ecotypes contained relatively higher saponin contents compared with African ecotypes, suggesting that *M. truncatula* ecotypes modify their secondary metabolism to acclimatize to their environments. In addition, tissue-specific saponin contents were also found between the shoot and the root tissues of the same ecotypes. For example, some saponins were observed in both the shoot and root tissues, whereas zanhic acid glycosides were found specifically in the shoot tissues, but not on the root tissues. On the other hand, bayogenin, hederagenin, and soyasaponin B glycosides were found predominantly in root tissues (Lei et al. 2019). The variation in saponin contents and types between root and shoot tissues suggests that an ecological roles for these tissue-specific saponin compounds in plant (Lei et al. 2019). For instance, root saponins hederagenin and bayogenin glycosides may protect against soil microbes, whereas, aerial saponins such as zanhic glycosides may act as herbivores deterrent (Lei et al. 2019) (Fig. 6.2).

With respect to legume saponins and antifungal activity, Martyniuk and Biały (2008) examined the antifungal activity of hederagenin and bayogenin glycosides isolated from *M. arabica* against *Cephalosporium gramineum*. In general, bayogenin glycosides exhibited stronger inhibitory effects on *C. gramineum* compared with hederagenin glycosides. In addition, the monodesmoside saponins with one sugar chain attached at the C3 position of bayogenin and hederagenin glycosides

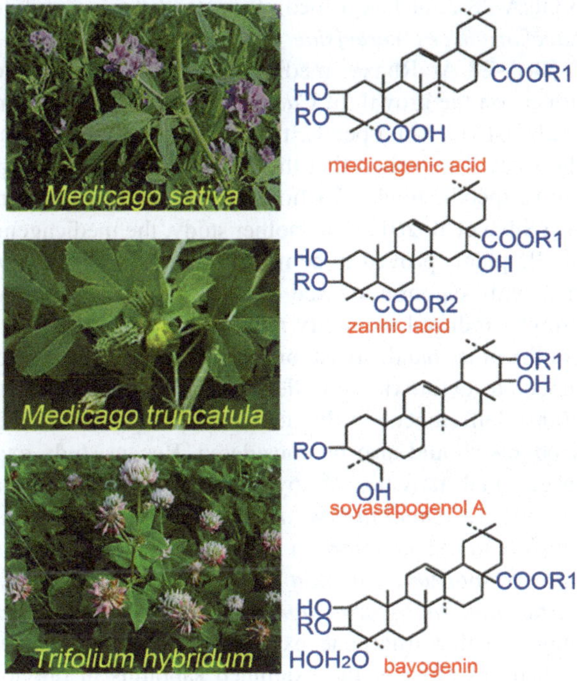

Fig. 6.2 Triterpene saponins isolated from different legume species

were much stronger against the growth of the fungus *C. gramineum* compared with bidesmoside saponins with two sugar chains linked at the C3 and C28 positions (Martyniuk and Biały 2008). Likewise, Goławska et al. (2012) studied the relationship between quantitative and qualitative variations of saponin contents in foliar tissues of European alfalfa cultivars namely, 'Sapko', 'Radius', 'Sitel' and 'Radius1' on the growth and development of pea aphid (*Acyrthosiphon pisum*). 'Radius', 'Sapko', and 'Sitel' cultivars possessed the three main saponin compounds, including zanhic acid tridesmoside, medicagenic acid and soyasapogenol B, whereas the 'Radius1' cultivar did not contained medicagenic acid and zanhic acid tridesmoside saponins. Total saponin content was highest in 'Radius' cultivar and lowest in 'Radius1' (Goławska et al. 2012). In addition, the aphid-infested plants exhibited higher saponin content than aphid-uninfected plants, regardless to the cultivar type. Consistently, the total number of aphid numbers were highest on 'Radius1' characterized by lower saponin contents. On the other hand 'Radius' cultivar with highest saponin contents showed lowest aphid numbers (Goławska et al. 2012). These results suggested a negative relationship between saponin content and aphid number, indicating that saponins in alfalfa plants have herbivore-induced defense, thus breeding strategies aiming to increase the levels of saponins in the foliage of infested alfalfa might be an efficient strategy to improve alfalfa resistance against aphid. In another study, 3-GlcA-28-AraRhaxyl-medicagenate saponin isolated from *M. truncatula* seed flour displayed a robust toxic activity against the rice weevil *Sitophilus oryzae*, a key pest of stored cereals (Da Silva et al. 2012). In addition,

3-GlcA-28-AraRhaxyl-medicagenate saponin inhibited the growth of the yeast *Saccharomyces cerevisiae* at concentrations higher than 100 μg/mL; however, 3-GlcA-28-AraRhaxyl-medicagenate saponin did not displayed any inhibitory effects on the growth of *Caenorhabditis elegans* worm or bacteria *E. coli* (Da Silva et al. 2012). . This specificity of GlcA-28-AraRhaxyl-medicagenate saponin against the weevil, indicated that this saponin can be potential application for pest control with a specific mode of action, rather than acting as non-specific detergent properties (Da Silva et al. 2012). In another study, the medicagenic acid saponins isolated from alfalfa plants proved to be most active against *Spodoptera littoralis*, and larval diet rich with saponins caused prolongation of the larval and pupal stages, retarded growth, reduced fecundity and fertility and increased mortality (Adel et al. 2000). On the other hand, soysaponogenol A, soysaponogenol B and hederagenin showed moderate cytotoxic activities, while soysaponogenol E was inactive (Adel et al. 2000). Differences in the activities of saponin acids indicate mutual synergism of saponins should also be considered. Recent study examined the antimicrobial and insecticidal activities of soy saponins against bacterial pathogens *Staphylococcus aureus*, *S. epidermiditis*, *Pseudomonas aeruginosa*, *Escherichia coli*, *Erwinia amylovora*, *Agrobacterium tumefaciens* and *E. carotovora*, fungi pathogens *Fusarium oxysporium*, *Candida albicans* and *Botrytis cinerea*, and insect *Tribolium castaneum*, *Rhyzopertha dominica* and *Sitophilus oryzae* (Allam et al. 2017). Results indicated that triterpene extracts exhibited an antibacterial activity which is not realistic as most of the examined saponins in previous reported clearly indicated that saponins doesn't possessed antibacterial activity, thus the observed results in soybean extract might be due to other components rather than triterpene saponins, especially the authors did not examined individual pure compounds, and they use crude extract method which is not realistic for specific activity (Abdelrahman et al. 2014). In addition, the antifungal activity of the soy triterpene saponins is in accordance with previous reports (Allam et al. 2017). Also authors reported that the toxicity of soy triterpene saponins was more effective against *Tribolium castaneum* insect than *Rhyzopertha dominica* and *Sitophilus oryzae*, indicating a specification of the triterpene bioactivity. Soy saponins can be also transformed into different saponin groups by fungi to reduce their toxicity and antifungal activity. A recent study examined the transformation rate of soy saponin into soyasapogenol B by different fungal isolates (Amin et al. 2013). Results indicated that *Aspergillus parasiticus* produced the highest yield 65% of soyasapogenol B after 72 h incubation at 33 °C (Amin et al. 2013).

6.2.2 Triterpenoid Saponins from the Genus *Camellia*

The genus *Camellia* contains 280 plant species, and the majority of them are located in tropical and subtropical Asian countries. Different *Camellia* sp. exhibited potential economic importance such as *C. sinensis* var. assamica, *C. sinensis* *C. reticulata*, *C. oleifera*, *C. sasanqua* and *C. japonica*. For instance, *C. sinensis* var. assamica and *C. sinensis* leaves are major tea producing materials and most popular beverages worldwide (Cui et al. 2018). In addition, *C. japonica*, *C. reticulata*, and *C. sasanqua* are well-known ornamental plants, while the seeds of *C. oleifera* are used for the production of edible oil (Cui et al. 2018). Up to date, most of the saponins have been isolated from the genus *Camellia* are pentacyclic triterpenoid saponins, and most of them are oleanane-type triterpenoid saponins (Cui et al. 2018). For example many pentacyclic triterpenoid saponin monomers have been isolated and characterized from the seeds, flowers, stems and roots of *C. sinensis* var. assamica, *C. oleifera*, *C. sinensis*, *C. sasanqua* and *C. japonica*. The isolated saponin compounds from different *Camellia* sp. were characterized by a sugar unit attached at C3 position and an acyl groups attached at C16, C21, C22, C23 and/or C28 of the saponin aglycone structure (Sagesaka et al. 1994; Lu et al. 2000; Murakami et al. 2000; Li et al. 2013; Uddin et al. 2014; Cui et al. 2018). Many research articles have been published regarding the *Camellia* saponins especially for the various methods of extractions and the biological properties of the crude saponins, including anti-microbial, anti-oxidant, and anti-inflammatory activities (Li et al. 2013; Uddin et al. 2014; Khan et al. 2018), however, most of them examined a complex saponin mixture, and the specific patterns of the structure-activity relationships are not confirmed.

6.2.2.1 Chemical Structure and Purification of Saponins from *Camellia* sp.

The dry *Camellia* seeds are grinded into fine powder, and the powder materials are defatted using *n*-hexane or petroleum ether to remove oils and fats. The defatted materials then extracted by classical techniques, such as reflux and maceration extractions or through ultrasonic and microwave-assisted extractions (Uddin et al. 2014; Yu and He 2018). In general, 50–80% MeOH or EtOH solutions are commonly used as extraction solvents, and the crude saponin extracts can be further purified using diethyl ether or acetone precipitation, separator-funnel partition, and/or column chromatography with reversed phase silica gel (Li et al. 2013; Myose et al. 2012; Zhang et al. 2015; Guo et al. 2018). After isolation of purified compound, Structure of saponin compounds can be elaborated by mass spectrometer (MS), infrared spectrometry (IR), ID or 2D nuclear magnetic resonance (NMR) including ^{13}C-NMR and ^{1}H-NMR, distortionless enhancement of polarization transfer (DEPT), 1H-1H correlated spectroscopy (COSY), nuclear overhauser effect spectroscopy (NOESY), total correlation spectroscopy (TOCSY), heteronuclear multiple bond coherence (HMBC), heteronuclear singular quantum coherence (HSQC), and rotating frame overhauser effect spectroscopy (ROESY) (Zhang et al. 2012; Fu et al. 2017; Guo et al. 2018). On the other hand, sugar moieties can be analyzed by GC-MS or HPLC after acid hydrolysis (Zong and Wang 2015; Guo et al. 2018).

6.2.2.2 Structure and Distribution of Triterpene Saponins from *Camellia* sp.

Theasaponins the main saponin compounds isolated from *Camellia* sp. are triterpenoid saponins consisted of sapogenin/glycone skeleton, oligosaccharide chains and organic acids, and these triterpene saponins are usually classified as oleanane-type pentacyclic triterpenoids (Cui et al. 2018). Patterns of theasaponin structures observed in different *Camellia* sp. indicated the effects of genetic relationships on the saponin structure, however, different saponin compounds were also isolated even from the same species, indicting the effects of the environmental factors such as temperature, humidity and soil fertility on the saponin structure. The sugar moieties are include glucose, galactose, glucuronic acid, arabinose, xylose and rhamnose. On the other hand, the basic carbon frame of oleanolic acid acts as the pentahedral nucleus of the polyhydrogen pin, and the spatial configurations of the rings are A/B-*trans*, B/C-*trans*, C/D-*trans*, and D/E-*cis* (Cui et al. 2018). While, the C12 and C13 positions form the unsaturated double bond acts as the nuclear parent. The majority of the theasaponins were isolated from the seeds of *Camellia* sp., including, theasaponin E2 methyl ester and oleiferasaponins (A, B, B, C and D) from the seeds of *C. oleifera*; theasaponins (A, B, C, E, F, G, and H), Assamsaponins (A and J), teaseedsaponins (A and L), floratheasaponin A, foliatheasaponins (I, II and III) and 21-O-angeloyltheasapogenol E3 from the seeds of *C. sinensis*; Camelliasaponins (A, B and C) from the seeds of *C. japonica* as well as Camelliasaponin B and C from the seeds of *C. oleifera*, *C. sinensis*, and *C. japonica* (Yoshikawa et al. 1994; Huang et al. 2005; Chen et al. 2010; Kuo et al. 2010; Myose et al. 2012; Zhang et al. 2012; Joshi et al. 2013; Li et al. 2013; Zhou et al. 2014; Yang et al. 2014; Zong and Wang 2015; Fu et al. 2017; Guo et al. 2018). These saponin compounds are usually oleanane-type triterpene saponins, and their structural diversity is derived from the vast array of the sapogenin skeleton and attached sugar chains. Further structure diversity can be also resulted from the presence of angeloyl, tigloyl, acetyl, 2-methylbutyryl, hexenoyl, isovaleryl, cinnamoyl and hydrocinnamoyl attached to the hydroxyl group at positions C16, C21, C22 and C28 of the aglycone skeleton. However, oligosaccharidic moieties can be commonly found as a D-glucuronopyranosyl or its methyl ester at C3 position, and substituted at position 2′ and 3′ positions by β-D-galactopyranosyl, β-D-glucopyranosyl, α-L-arabinopyranosyl, β-D-xylopyranosyl and α-L-rhamnopyranosyl has been reported. It worth nothing that, D-glucuronic acid methyl ester was found only in several saponins from the seeds of *C. oleifera*. In Table 6.1, we summarized recent saponins isolated and identified from different *Camellia* sp.

6.3 Steroidal Saponins

Steroidal saponins are mainly found among monocot plants, including *Allium*, *Aster* and *Asparagus* plants (Abdelrahman et al. 2014). The wide use of these plants in traditional medicines was mainly attributed to the rich amount of sulfur and saponin-related compounds (Lanzotti 2005). Although many studies have been conducted in saponin isolation and identification, the research related to steroidal saponins is much less than triterpene saponins in terms of number of isolated compounds and examined biological activities. The extraction of steroidal saponin compounds from plant materials can be achieved through using the traditional MeOH aqueous solvents. However, it's more efficient to use gradient solvent systems with different

Table 6.1 List of the some triterpene saponins isolated from *Camellia* sp.

Saponin name	*Camellia* sp.	Molecular formula	Reference
Theasaponin A_1	*C. sinensis*	$C_{57}H_{90}O_{26}$	Morikawa et al. (2006)
Theasaponin A_2	*C. sinensis*	$C_{59}H_{92}O_{27}$	Morikawa et al. (2006)
Theasaponin A_3	*C. sinensis*	$C_{61}H_{94}O_{28}$	Morikawa et al. (2006)
Theasaponin A_4	*C. sinensis*	$C_{58}H_{92}O_{27}$	Yoshikawa et al. (2007)
Theasaponin A_5	*C. sinensis*	$C_{60}H_{94}O_{28}$	Yoshikawa et al. (2007)
Theasaponin A_6	*C. sinensis*	$C_{60}H_{94}O_{28}$	Morikawa et al. (2007)
Theasaponin A_7	*C. sinensis*	$C_{62}H_{96}O_{29}$	Morikawa et al. (2007)
Theasaponin A_8	*C. sinensis*	$C_{61}H_{94}O_{28}$	Li et al. (2013)
Theasaponin A_9	*C. sinensis*	$C_{59}H_{92}O_{27}$	Li et al. (2013)
Theasaponin B_5	*C. sinensis*	$C_{57}H_{90}O_{25}$	Morikawa et al. (2007)
Theasaponin C_1	*C. sinensis*	$C_{57}H_{90}O_{25}$	Yoshikawa et al. (2007)
Assamsaponin A	*C. sinensis*	$C_{57}H_{88}O_{25}$	Murakami et al. (1999)
Assamsaponin B (Tea saponin S1)	*C. sinensis*	$C_{61}H_{92}O_{28}$	Joshi et al. (2013); Murakami et al. (1999)
Assamsaponin C (Tea saponin S4)	*C. sinensis*	$C_{61}H_{92}O_{28}$	Joshi et al. (2013); Murakami et al. (1999)
Assamsaponin D	*C. sinensis*	$C_{59}H_{92}O_{27}$	Murakami et al. (1999)
Assamsaponin E	*C. sinensis*	$C_{57}H_{88}O_{25}$	Murakami et al. (1999)
Assamsaponin F	*C. sinensis*	$C_{62}H_{94}O_{29}$	Murakami et al. (2000)
Assamsaponin G	*C. sinensis*	$C_{60}H_{92}O_{28}$	Murakami et al. (2000)
Assamsaponin H	*C. sinensis*	$C_{60}H_{92}O_{28}$	Murakami et al. (2000)
Assamsaponin I	*C. sinensis*	$C_{60}H_{92}O_{28}$	Murakami et al. (2000)
Assamsaponin J	*C. sinensis*	$C_{53}H_{86}O_{24}$	Murakami et al. (2000)
Teaseedsaponin A	*C. sinensis*	$C_{62}H_{96}O_{28}$	Myose et al. (2012)
Teaseedsaponin B	*C. sinensis*	$C_{62}H_{96}O_{28}$	Myose et al. (2012)
Teaseedsaponin C	*C. sinensis*	$C_{58}H_{92}O_{25}$	Myose et al. (2012)
Teaseedsaponin D	*C. sinensis*	$C_{63}H_{98}O_{28}$	Myose et al. (2012)
Teaseedsaponin E	*C. sinensis*	$C_{62}H_{96}O_{27}$	Myose et al. (2012)
Teaseedsaponin F	*C. sinensis*	$C_{62}H_{98}O_{27}$	Myose et al. (2012)
Teaseedsaponin G	*C. sinensis*	$C_{58}H_{90}O_{25}$	Myose et al. (2012)
Teaseedsaponin H	*C. sinensis*	$C_{61}H_{94}O_{28}$	Myose et al. (2012)
Teaseedsaponin I	*C. sinensis*	$C_{63}H_{96}O_{28}$	Myose et al. (2012)
Teaseedsaponin J	*C. sinensis*	$C_{62}H_{94}O_{27}$	Myose et al. (2012)
Teaseedsaponin K	*C. sinensis*	$C_{62}H_{94}O_{27}$	Myose et al. (2012)
Teaseedsaponin L	*C. sinensis*	$C_{62}H_{94}O_{28}$	Myose et al. (2012)
Floratheasaponin A	*C. sinensis*	$C_{59}H_{92}O_{26}$	Yoshikawa et al. (2005)
Foliatheasaponin I	*C. sinensis*	$C_{61}H_{94}O_{27}$	Li et al. (2008)
Foliatheasaponin III	*C. sinensis*	$C_{61}H_{94}O_{27}$	Morikawa et al. (2007)
21-O-Angeloyltheasapogenol E3	*C. sinensis*	$C_{37}H_{56}O_8$	Yang et al. (2014)
Camelliasaponin A1	*C. japonica*	$C_{58}H_{92}O_{25}$	Yoshikawa et al. (1996)
Camelliasaponin A2	*C. japonica*	$C_{58}H_{92}O_{25}$	Yoshikawa et al. (1996)

(continued)

Table 6.1 (continued)

Saponin name	*Camellia* sp.	Molecular formula	Reference
Camelliasaponin B1	*C. japonica*	$C_{58}H_{90}O_{26}$	Kuo et al. (2010)
Camelliasaponin B2	*C. japonica*	$C_{58}H_{90}O_{26}$	Yoshikawa et al. (1994)
Camelliasaponin C1	*C. japonica*	$C_{58}H_{92}O_{26}$	Yoshikawa et al. (1994)
Camelliasaponin C2	*C. japonica*	$C_{58}H_{92}O_{26}$	Yoshikawa et al. (1994)
Oleiferasaponin A1	*C. oleifera*	$C_{59}H_{92}O_{26}$	Zhang et al. (2012)
Camellioside A	*C. japonica*	$C_{53}H_{84}O_{24}$	Yoshikawa et al. (2007)
Camellioside B	*C. japonica*	$C_{55}H_{86}O_{25}$	Yoshikawa et al. (2007)
Camellioside C	*C. japonica*	$C_{53}H_{82}O_{23}$	Yoshikawa et al. (2007)
Camellioside D	*C. japonica*	$C_{54}H_{88}O_{24}$	Yoshikawa et al. (2008)
Chakasaponin I	*C. sinensis*	$C_{59}H_{92}O_{26}$	Yoshikawa et al. (2008)
Chakasaponin II	*C. sinensis*	$C_{62}H_{96}O_{27}$	Yoshikawa et al. (2008)
Chakasaponin III	*C. sinensis*	$C_{59}H_{92}O_{27}$	Yoshikawa et al. (2008)
Chakasaponin IV	*C. sinensis*	$C_{57}H_{90}O_{25}$	Matsuda et al. (2012)
Chakasaponin V	*C. sinensis*	$C_{63}H_{98}O_{27}$	Yoshikawa et al. (2008)
Chakasaponin VI	*C. sinensis*	$C_{59}H_{92}O_{27}$	Yoshikawa et al. (2008)
Yuchasaponin A	*C. oleifera*	$C_{64}H_{100}O_{28}$	Sugimoto et al. (2009)
Yuchasaponin B	*C. oleifera*	$C_{64}H_{100}O_{28}$	Sugimoto et al. (2009)
Yuchasaponin C	*C. oleifera*	$C_{64}H_{100}O_{28}$	Sugimoto et al. (2009)
Yuchasaponin D	*C. oleifera*	$C_{64}H_{100}O_{27}$	Sugimoto et al. (2009)
Jegosaponin B	*C. oleifera*	$C_{61}H_{96}O_{27}$	Sugimoto et al. (2009)
Sasanquasaponins I	*C. sasanqua*	$C_{60}H_{96}O_{26}$	Nakamura et al. (2012); Sugimoto et al. (2009)
Sasanquasaponins II	*C. sasanqua*	$C_{59}H_{94}O_{26}$	Nakamura et al. (2012); Sugimoto et al. (2009)
Sasanquasaponins III	*C. sasanqua*	$C_{59}H_{94}O_{26}$	Nakamura et al. (2012); Sugimoto et al. (2009)
Sasanquasaponins IV	*C. sasanqua*	$C_{59}H_{94}O_{26}$	Matsuda et al. (2010); Fujimoto et al. (2012)
Sasanquasaponins V	*C. sasanqua*	$C_{59}H_{96}O_{25}$	Matsuda et al. (2010); Fujimoto et al. (2012)
Sanchakasaponins A	*C. japonica*	$C_{53}H_{84}O_{23}$	Nakamura et al. (2012); Sugimoto et al. (2009)
Sanchakasaponins B	*C. japonica*	$C_{59}H_{96}O_{26}$	Nakamura et al. (2012); Sugimoto et al. (2009)
Sanchakasaponins C	*C. japonica*	$C_{64}H_{100}O_{28}$	Nakamura et al. (2012); Sugimoto et al. (2009)
Sanchakasaponins D	*C. japonica*	$C_{64}H_{100}O_{28}$	Nakamura et al. (2012); Sugimoto et al. (2009)
Sanchakasaponins E	*C. japonica*	$C_{61}H_{96}O_{27}$	Nakamura et al. (2012); Sugimoto et al. (2009)
Sanchakasaponins F	*C. japonica*	$C_{64}H_{100}O_{28}$	Nakamura et al. (2012) Sugimoto et al. (2009)
Sanchakasaponins G	*C. japonica*	$C_{59}H_{94}O_{25}$	Nakamura et al. (2012); Sugimoto et al. (2009)

(continued)

Table 6.1 (continued)

Saponin name	*Camellia* sp.	Molecular formula	Reference
Sanchakasaponins H	*C. japonica*	$C_{59}H_{94}O_{25}$	Nakamura et al. (2012); Sugimoto et al. (2009)
Maetenoside B	*C. japonica*	$C_{59}H_{94}O_{25}$	Nakamura et al. (2012); Sugimoto et al. (2009)
Ternstoemiaside C	*C. japonica*	$C_{54}H_{88}O_{24}$	Nakamura et al. (2012); Sugimoto et al. (2009)
Primulagenin A-S8	*C. sasanqua*	$C_{54}H_{88}O_{23}$	Fujimoto et al. (2012); Matsuda et al. (2010)
Floraassamsaponins I	*C. sinensis* var. *assamica*	$C_{66}H_{104}O_{31}$	Ohta et al. (2015)
Floraassamsaponins II	*C. sinensis* var. *assamica*	$C_{66}H_{104}O_{31}$	Ohta et al. (2015)
Floraassamsaponins III	*C. sinensis* var. *assamica*	$C_{60}H_{94}O_{27}$	Ohta et al. (2015)
Floraassamsaponins IV	*C. sinensis* var. *assamica*	$C_{60}H_{94}O_{27}$	Ohta et al. (2015)
Floraassamsaponins V	*C. sinensis* var. *assamica*	$C_{60}H_{94}O_{27}$	Ohta et al. (2015)
Floraassamsaponins VI	*C. sinensis* var. *assamica*	$C_{60}H_{94}O_{26}$	Ohta et al. (2015)
Floraassamsaponins VII	*C. sinensis* var. *assamica*	$C_{60}H_{94}O_{26}$	Ohta et al. (2015)
Floraassamsaponins VIII	*C. japonica*	$C_{60}H_{94}O_{26}$	Ohta et al. (2015)
TR-Saponin A	*C. sinensis* var. *assamica*	$C_{53}H_{80}O_{20}$	Lu et al. (2000)
TR-Saponin B	*C. sinensis* var. *assamica*	$C_{53}H_{82}O_{20}$	Lu et al. (2000)
TR-Saponin C	*C. sinensis* var. *assamica*	$C_{55}H_{84}O_{21}$	Lu et al. (2000)
Rogchaponin R1	*C. sinensis*	$C_{52}H_{80}O_{21}$	Varughese et al. (2011)
Rogchaponin R2	*C. sinensis*	$C_{52}H_{78}O_{21}$	Varughese et al. (2011)
Rogchaponin R3	*C. sinensis*	$C_{50}H_{76}O_{19}$	Varughese et al. (2011)
Rogchaponin R4	*C. sinensis*	$C_{53}H_{82}O_{20}$	Varughese et al. (2011)
Rogchaponin R5	*C. sinensis*	$C_{51}H_{78}O_{19}$	Varughese et al. (2011)
Rogchaponin R6	*C. sinensis*	$C_{53}H_{80}O_{20}$	Varughese et al. (2011)
Rogchaponin R7	*C. sinensis*	$C_{54}H_{82}O_{21}$	Varughese et al. (2011)
Rogchaponin R8	*C. sinensis*	$C_{51}H_{80}O_{18}$	Varughese et al. (2011)
Rogchaponin R9	*C. sinensis*	$C_{53}H_{82}O_{19}$	Varughese et al. (2011)
Rogchaponin R10	*C. sinensis*	$C_{54}H_{84}O_{19}$	Varughese et al. (2011)
Oleiferoside A	*C. oleifera*	$C_{63}H_{96}O_{29}$	Li et al. (2014)
Oleiferoside B	*C. oleifera*	$C_{63}H_{98}O_{29}$	Li et al. (2014)
Oleiferoside C	*C. oleifera*	$C_{58}H_{92}O_{26}$	Li et al. (2014)
Oleiferoside D	*C. oleifera*	$C_{64}H_{98}O_{29}$	Li et al. (2014)

(continued)

Table 6.1 (continued)

Saponin name	*Camellia* sp.	Molecular formula	Reference
Oleiferoside E	*C. oleifera*	$C_{50}H_{80}O_{19}$	Li et al. (2014)
Oleiferoside F	*C. oleifera*	$C_{54}H_{82}O_{19}$	Li et al. (2014)
Oleiferoside G	*C. oleifera*	$C_{60}H_{92}O_{23}$	Li et al. (2014)
Oleiferoside H	*C. oleifera*	$C_{60}H_{92}O_{23}$	Li et al. (2014)
Oleiferoside I	*C. oleifera*	$C_{51}H_{78}O_{19}$	Li et al. (2015)
Oleiferoside J	*C. oleifera*	$C_{63}H_{98}O_{28}$	Li et al. (2015)
Oleiferoside K	*C. oleifera*	$C_{63}H_{100}O_{28}$	Li et al. (2015)
Oleiferoside L	*C. oleifera*	$C_{63}H_{96}O_{28}$	Li et al. (2015)
Oleiferoside M	*C. oleifera*	$C_{63}H_{100}O_{27}$	Li et al. (2015)
Oleiferoside N	*C. oleifera*	$C_{58}H_{88}O_{23}$	Yang et al. (2015)
Oleiferoside O	*C. oleifera*	$C_{58}H_{90}O_{23}$	Yang et al. (2015)
Oleiferoside P	*C. oleifera*	$C_{55}H_{82}O_{21}$	Wu et al. (2015)
Oleiferoside Q	*C. oleifera*	$C_{55}H_{84}O_{21}$	Wu et al. (2015)
Oleiferoside R	*C. oleifera*	$C_{53}H_{82}O_{19}$	Wu et al. (2015)
Oleiferoside S	*C. oleifera*	$C_{53}H_{82}O_{20}$	Wu et al. (2015)
Oleiferoside T	*C. oleifera*	$C_{53}H_{80}O_{19}$	Wu et al. (2015)
Oleiferoside U	*C. oleifera*	$C_{52}H_{82}O_{23}$	Zhang et al. (2016)
Oleiferoside V	*C. oleifera*	$C_{55}H_{82}O_{23}$	Zhang et al. (2016)
Oleiferoside W	*C. oleifera*	$C_{58}H_{92}O_{26}$	Wu et al. (2018)
Oleiferasaponin B1	*C. oleifera*	$C_{58}H_{90}O_{26}$	Zhou et al. (2014)
Oleiferasaponin B2	*C. oleifera*	$C_{61}H_{90}O_{24}$	Zhou et al. (2014)
Oleiferasaponin C1	*C. oleifera*	$C_{59}H_{92}O_{26}$	Zong and Wang (2015)
Oleiferasaponin C2	*C. oleifera*	$C_{60}H_{96}O_{26}$	Zong and Wang (2015)
Oleiferasaponin C4	*C. oleifera*	$C_{60}H_{94}O_{27}$	Zong et al. (2016)
Oleiferasaponin C5	*C. oleifera*	$C_{54}H_{84}O_{22}$	Zong et al. (2016)
Oleiferasaponin C6	*C. oleifera*	$C_{65}H_{94}O_{28}$	Zong et al. (2016)
Oleiferasaponin D1	*C. oleifera*	$C_{58}H_{91}O_{25}$	Fu et al. (2017)
Oleiferasaponin D2	*C. oleifera*	$C_{58}H_{89}O_{26}$	Fu et al. (2017)
Oleiferasaponin D3	*C. oleifera*	$C_{58}H_{89}O_{26}$	Fu et al. (2017)
Oleiferasaponin D4	*C. oleifera*	$C_{58}H_{91}O_{26}$	Fu et al. (2017)
Oleiferasaponin D5	*C. oleifera*	$C_{59}H_{91}O_{27}$	Fu et al. (2017)
Theasaponin E2 methyl ester	*C. oleifera*	$C_{60}H_{92}O_{27}$	Chen et al. (2010)
Camelliaolean A	*C. japonica*	$C_{31}H_{50}O_{6}$	Uddin et al. (2014)
Camelliaolean B	*C. japonica*	$C_{30}H_{48}O_{6}$	Uddin et al. (2014)

polarities to remove other non-essential compounds. For instance, the dry plant material can be extracted with hexane to remove oils, followed by $CHCl_3$ to remove low molecular weight compounds, then crude saponin fraction can be obtained by using $CHCl_3$:MeOH (9:1, v:v) and MeOH. To remove sugars and nucleotides, the methanol extract dissolved in water and portioned with n-butanol. Water phase discarded and n-butanol phase then collected and dried under pressure and dissolved in 80–90% MeOH and subjected to column chromatography or TLC to isolate different saponin compounds (Mostafa et al. 2013). For structure elucidation, HRFABMS and advanced 1D and 2D NMR experiments are mainly used. The 2D NMR method simplified the structure clarification of organic compounds because it can show the interactions between nuclei (Lanzotti 2005). The interpretation of 2D NMR spectra is frequently straightforward, and results can be correlated through homonuclear coupling, including COSY, HOHAHA or TOCSY, ROESY, and heteronuclear coupling using ^{1}H-detected experiments such as HMQC or HSQC, and HMBC (Lanzotti 2005).

6.3.1 Steroidal Saponins from Monocotyledonous Plants

The genus *Allium* comprise approximately 850 species, which are widely distributed in nature especially in the northern hemisphere (Abdelrahman et al. 2016; Abdelrahman et al. 2017a, b). Approximately, more than 130 spirostanol saponin glycosides have been isolated and identified from various *Allium* species (Ikeda et al. 2000). In general, most of the *Allium* spirostane-type saponins possessed monodesmodic with one sugar chain attached at C3 position of the aglycone. However some cases, the sugar chain might attached at C1 position such as in alliospirosides A-D, C24 position such as in chinenoside VI, karatavioside F, and anzuroside, or even at C27 position such as in tuberoside L (Kravets et al. 1986a, b; Jiang et al. 1998; Vollerner et al. 1984, 1989; Sang et al. 2001). Also around 140 furostanol saponin glycosides have been identified in the genus *Allium*. Furostanol-type saponins possess either a *trans* or a *cis* fusion between ring A and B, or a double bond between C5 and C5 position, leading to 5α, 5β or Δ^{5} series. Several frustanol saponins have been isolated from *Allium* species, inducing ascalonicoside B, ceparoside C, chinenoside II (Fattorusso et al. 2002; Yuan et al. 2009; Peng et al. 1996). The early study by Kawashima et al. (1991) using chemical investigation of the bulbs of *A. giganteum* and *A. aflatunense* has resulted in the isolation and identification of two new steroidal saponins. The chemical structure of the new isolated spirostanol saponin from *A. giganteum* was identified as (24S,25R)-5α-spirostan-2α,3β,5α,6β,24-pentaol 24-O-β-d-glucopyranoside, while the new isolated spirostanol saponin isolated from *A. aflatunense* was identified as (25R)-5α-spirostan-2α,3β,5α,6α-tetraol 2-O-β-d-glucopyranoside (Kawashima et al. 1991). The MeOH extraction of the bulbs of *A. aflatunense* and *A. giganteum* was subjected to silica gel and DIAION HP-20 column chromatography, and reversed phase HPLC. Then, the chemical structure was further conformed using ^{1}H NMR and ^{13}C NMR spectroscopy combined with 2D NOESY spectrum (Kawashima et al. 1991). Later on the same research group (Sashida et al. 1991) was able to isolate and identify three new steroidal saponin compounds, namely(25R)-3-O-benzoyl-5-α-spirostan-2α, 3β, 5α, 6β-tetraol 2-O-β-D-glucopyranoside; (25R)-3-O-acetyl-5-α-spirostan-2α, 3β, 5α, 6β-tetraol 2-O-β-D-glucopyranoside and (25R)-5α-spirostan-2α, 3β, 5α, 6β-tetraol 2-O-β-D-glucopyranoside from the bulb of *A. giganteum*. In another research study, the structure elucidation of new steroidal saponin compound isolated from the flower tissue of *A. leucanthum* C. was confirmed NMR and HRESIMS spectrometry analyses (Mskhiladze et al. 2008). The new saponin compound was named leucospiroside A and its chemical structure has been identified as (25R)-5α-spirostane-2α,3β,6βtriol 3-O-β-glucopyranosyl-(1→3)-β-glucopyranosyl-(1→2)-[β-glucopyranosyl-(1→3)]-βglucopyranosyl-(1→4)-β-galactopyranoside (Mskhiladze et al. 2008). In addition, the cytotoxic activities of the new isolated compound, and three kwon compound as well as the crude saponin factions were examined for their in vitro cytotoxic activity against lung cancer cell line (A549) and colon cancer cell line (DLD-1). In general, the new leucospiroside A saponin and the know saponin compounds exhibited strong cytotoxic activity compared with crude saponin fractions (Mskhiladze et al. 2008). Likewise, the phytochemical analyses of the saponin extract isolated from *A. porrum*, resulted in the isolation and identification eight saponin compounds, and out of them, four saponins (compound 5–8) were identified as novel compounds. The new isolated saponin compound 5 and 6, showed the same tetrasaccharide moiety with saponin compound 1 and 3, but they also possessed unusual spirostane aglycones, namely 12-ketoporrigenin and 2,12-diketoporrigenin (named porrigenin C), respectively. In addition, the new compound 7 and 8 were identified as rare cholestane bidesmosides with di- and trisaccharide residues attached to a polyhydroxy cholesterol aglycone, respectively. The in vitro cytotoxic activity of all the eight saponin compounds was evaluated WEHI 164 and J774 cell lines, demonstrating that compound 1, 2, and 6 displayed the highest cytotoxic activities (Fattorusso et al. 2000). Similarly, the cytotoxic activity of crude saponin faction isolated from *A. chinense* using ethanol extraction and further purified with the D101 macroporous adsorption resin approach against B16 melanoma and 4T1 breast carcinoma cell lines was carried out. Saponin-treated cell lines exhibited morphological changes, and steroidal saponin treatments inhibited cell migration and colony formation and induced cell death in B16 and 4T1 cells in a concentration-dependent manner (Yu et al. 2015). In a recent study, a steroidal saponin, named Cepa2, was isolated from the dry roots of shallot (*A. cepa* L. Aggregatum group), and chemical structure was confirmed using 2D NMR. The ^{1}H NMR and ^{13}C NMR data revealed that the isolated Cepa2 compound is identical to the previously identified alliospiroside A (Abdelrahman et al. 2017a, b). In vitro cytotoxic activity of Cepa2/ alliospiroside A against P3U1 myeloma cancer cell line showed 91.13% reduction in P3U1 cell viability after12 h (Abdelrahman et al. 2017a, b). In addition, the reduction of cell viability was consisted with the increase in ROS levels P3U1-treated cells compared with untreated ones. The scanning-electron microscope demonstrated a clear apoptosis of the Cepa2-treated P3U1 cells in a time course and dose-dependent effect (Abdelrahman et al. 2017a, b). Interestingly, the same research group also reported that the addition of chromosome 2A from shallot into *A. fistulosum* enhanced the biosynthesis and accumulation of alliospiroside A, and improved their antifungal properties against Fusarium pathogens (Abdelrahman et al. 2017a, b). Additionally, Teshima et al. (2013), were able to isolate alliosprisode from shallots, revealing in vitro antifungal activity against different fungi pathogens. These results indicated that alliospiroside A isolated from shallot roots, is potential compound for disease resistance and pharmacological industries. In large screening experiments, 22 C-27 steroidal saponin compounds and 6 steroidal sapogenin compounds isolated from different monocotyledonous plants were investigated for their antifungal activities against *Candida glabrata, Candida albicans*, *Candida krusei*, *Aspergillus fumigatus*, and *Cryptococcus neoformans* (Yang et al. 2006). The obtained results indicated that the antifungal activities of the steroidal saponin compounds were highly interlinked with the nature of their aglycone units and the structure and number of monosaccharide moieties attached to the aglycone skeleton (Yang et al. 2006). Among all tested steroidal saponins, four tigogenin saponins with four or five monosaccharide chains exhibited significant activity against *A. fumigatus* and *C. neoformans*, comparable to the standard antifungal amphotericin B, indicating that the C-27 steroidal saponins may be considered potential antifungal leads for further medicinal and pharmaceutical industries (Yang et al. 2006). Also minutoside saponins and sapogenins, neoagigenin and alliogenin, isolated from the bulbs of *A. minutiflorum* displayed antimicrobial activity against different air-borne and soil-borne pathogenic fungi (Barile et al., 2007). A comparative analysis of steroidal saponin antifungal activities against wide range of crop pathogens was evaluated based on effective dose (EC_{50}) methods (Trdá et al. 2019). Results indicated that aescin saponin showed strongest antifungal activity compared with other saponin compounds. However, the antifungal effects of aescin could be inverted by ergosterol application, indicating that aescin saponin can interfere with sterols in the fungal cell walls. In addition, the induction of defense response through aescin treatments in two different pathosystems was examined. Results indicated that, aescin activated *B. napus* defense through the activation of the salicylic acid (SA)-dependent pathway and oxidative burst against *Leptosphaeria maculans* (Trdá et al. 2019). Whereas, aescin inhibited *Pseudomonas syringae* pv tomato DC3000 colonization of *A. thaliana* through also the activation of SA-dependent immune systems, but without direct antibacterial activity (Trdá et al. 2019). The above results suggested that, aescin not only exhibited potent antifungal properties but also can activate plant immunity in two different plant species through SA-dependent pathway.

6.4 Conclusion

The triterpene and steroidal saponin compounds have been extensity studied with respect to their chemical structure and biological activities, including antimicrobial, anticancer and anti-inflammatory and others. However, the genetic differences and the downstream biosynthesis or regulatory genes involved in saponin biosynthesis and structure diversity is still widely unknown. Thus future studies using omics technology might provide useful information for the saponin biosynthesis pathway and subsequently enabled their application in plant biotechnology for crop disease resistance.

References

Abdelrahman M, Sho H, Ito S-I, Yamauchi N, Shigyo M (2014) Compartmentation and localization of bioactive metabolites in different organs of *Allium roylei*. Biosci Biotechnol Biochem 78:1112–1122

Abdelrahman M, Abdel-Motaal F, El-Sayed M, Jogaiah S, Shigyo M, Ito S-I, Tran LP (2016) Dissection of *Trichoderma longibrachiatum*-induced defense in onion (*Allium cepa* L.) against *Fusarium oxysporum* f. sp. cepa by target metabolite profiling. Plant Sci 246:128–138

Abdelrahman M et al (2017a) RNA-sequencing-based transcriptome and biochemical analyses of steroidal saponin pathway in a complete set of *Allium fistulosum—A. cepa* monosomic addition lines. PLoS One 12:e0181784

Abdelrahman M, Mahmoud HYAH, El-Sayed M, Tanaka S, Tran LP (2017b) Isolation and characterization of Cepa2, a natural alliospiroside A, from shallot (*Allium cepa* L. Aggregatum group) with anticancer activity. Plant Physiol Biochem 116:167–173

Abed El Aziz MM, Ashou RAS, Melad AG (2019) A review on saponins from medicinal plants: chemistry, isolation, and determination. J Nanomed Res 7:282–288

Abugabr Elhag HEE, Naila A, Ajit A, Abdul Aziz B, Sulaiman AZ (2018) Sequential extraction of saponins from *Eurycoma longifolia* roots by water extraction and ultrasound-assisted extraction. Mater Today Proc 5:21672–21681

Adel MM, Sehnal F, Jurzysta M (2000) Effects of alfalfa saponins on the moth *Spodoptera littoralis*. J Chem Ecol 26:1065–1078

Allam NG, Abou-Taleb HK, Aboobia MM (2017) Antimicrobial and pesticidal activities of soya saponin. J Basic Environ Sci 4:262–267

Amin HA, Hassan YM, Yehia SM (2013) Biotransformation of soybean saponin to soyasapogenol B by *Aspergillus parasiticus*. Egypt Pharm J 12:40–45

Bialy Z, Jurzysta M, Oleszek W, Piacente S, Pizza C (1999) Saponins in alfalfa (*Medicago sativa* L.) root and their structural elucidation. J Agric Food Chem 47:3185–3192

Challinor VL, De Voss JJ (2013) Open-chain steroidal glycosides, a diverse class of plant saponins. Nat Prod Rep 30:429–454

Challinor VL, Parsons PG, Chap S, White EF, Blanchfield JT, Lehmann RP, De Voss JJ (2012) Steroidal saponins from the roots of *Smilax* sp.: structure and bioactivity. Steroids 77:504–511

Chen JH, Wu HY, Liau BC, Chang CMJ, Jong TT, Wu LC (2010) Identification and evaluation of antioxidants defatted *Camellia oleifera* seeds by isopropanol salting-out pretreatment. Food Chem 121:1246–1254

Cui C, Zong J, Sun Y, Zhang L, Ho C-T, Wan X, Hou R (2018) Triterpenoid saponins from the genus *Camellia*: structures, biological activities, and molecular simulation for structure–activity relationship. Food Funct 9:3069–3091

Da Silva P, Eyraud V, Carre-Pierrat M, Sivignon C, Rahioui I, Royer C et al (2012) High toxicity and specificity of the saponin 3-GlcA-28-AraRhaxyl-medicagenate, from *Medicago truncatula* seeds, for *Sitophilus oryzae*. BMC Chem Biol 12:3

Fanani MZ, Fukushima EO, Sawai S, Tang J, Ishimori M, Sudo H, Ohyama K, Seki H, Saito K, Muranaka T (2019) Molecular basis of C-30 product regioselectivity of legume oxidases involved in high-value triterpenoid biosynthesis. Front Plant Sci 10:1520

Fattorusso E, Lanzotti V, Taglialatela-Scafati O, Di Rosa M, Ianaro A (2000) Cytotoxic saponins from bulbs of *Allium porrum* L. J Agric Food Chem 48:3455–3462

Fattorusso E, Iorizzi M, Lanzotti V (2002) Chemical composition of shallot (*Allium ascalonicum* Hort.). J Agric Food Chem 50:5686–5690

Fu HZ, Wan KH, Yan QW, Zhou GP, Feng TT, Dai M, Zhong RJ (2017) Cytotoxic triterpenoid saponins from the defatted seeds of *Camellia oleifera* Abel. J Asian Nat Prod Res 20:412–422

Fujimoto K, Nakamura S, Nakashima S, Matsumoto T, Uno K, Ohta T, Miura T, Matsuda H, Yoshikawa M (2012) Medicinal flowers. XXXV. Nor-oleanane-type and acylated oleanane-type triterpene saponins from the flower buds of Chinese *Camellia japonica* and their inhibitory effects on melanogenesis. Chem Pharm Bull 60:1188–1194

Geisler K, Hughes RK, Sainsbury F, Lomonossoff GP, Rejzek M, Fairhurst S et al (2013) Biochemical analysis of a multifunctional cytochrome P450 (CYP51) enzyme required for synthesis of antimicrobial triterpenes in plants. Proc Natl Acad Sci USA 110:35

Goławska S, Łukasik I, Wójcicka A, Sytykiewicz H (2012) Relationship between saponin content in alfalfa and aphid development. Acta Biol Crac Ser Bot 54:39–46

Guo N, Tong T, Ren N, Tu Y, Li B (2018) Saponins from seeds of genus *Camellia*: Phytochemistry and bioactivity. Phytochemistry 149:42–55

Huang Q, Shao L, He M, Chen H, Liu D, Luo Y, Dai Y (2005) Inhibitory effects of sasanquasaponin on over-expression of ICAM-1 and on enhancement of capillary permeability induced by burns in rats. Burns 31:637–642

Huhman DV, Berhow MA, Sumner LW (2005) Quantification of saponins in aerial and subterranean tissues of *Medicago truncatula*. J Agric Food Chem 53:1914–1920

Ikeda T, Tsumagari H, Nohara T (2000) Steroidal oligoglycosides from the seeds of *Allium tuberosum*. Chem Pharm Bull 48(3):362–365

Jiang Y, Wang N-L, Yao X-S et al (1998) Structural elucidation of the anticoagulation and anticancer constituents from *Allium chinense*. Yao Xue Xue Bao 33:355–361

Joshi R, Sood S, Dogra P, Mahendru M, Kumar D, Bhangalia S, Pal HC, Kumar N, Bhushan S, Gulati A (2013) In vitro cytotoxicity, antimicrobial, and metal-chelating activity of triterpene saponins from tea seed grown in Kangra valley, India. Med Chem Res 22:4030–4038

Kapusta I, Janda B, Stochmal A, Oleszek W (2005a) Determination of saponins in aerial parts of barrel medic (*Medicago truncatula*) by liquid chromatography-electrospray ionization/mass spectrometry. J Agric Food Chem 53:7654–7660

Kapusta I, Stochmal A, Perrone A, Piacente S, Pizza C, Oleszek W (2005b) Triterpene saponins from barrel medic (*Medicago truncatula)* aerial parts. J Agric Food Chem 53:2164–2170

Kawashima K, Mimaki Y, Sashida Y (1991) Steroidal saponins from *Allium giganteum* and *A. aflatunense*. Phytochemistry 30:3063–3067

Khan MI, Ahhmed A, Shin JH, Baek JS, Kim MY, Kim JD (2018) Green tea seed isolated saponins exerts antibacterial effects against various strains of gram positive and gram negative bacteria, a comprehensive study *in vitro* and in *vivo*. Evid Based Complement Alternat Med 2018:3486106

Kräutler B, Sahu NP, Banerjee S, Mondal NB, Mandal D (2008) Steroidal saponin. In: Kinghorn AD, Columbus OH, Falk H, Kobayashi LJ (eds) Fortschritte der chemie organischer naturstoffe/ progress in the chemistry of organic natural products. Springer, Vienna, pp 45–141

Kravets SD, Vollerner YS, Gorovits MB et al (1986a) Steroids of the spirostan and furostan series from plants of the genus *Allium* XXI. Structure of alliospiroside A and alliofuroside A from *Allium cepa*. Chem Nat Comp 22:174–181

Kravets SD, Vollerner YS, Gorovits MB et al (1986b) Steroids of the spirostan and furostan series from plants of the genus *Allium* XXII. The structure of alliospiroside B from *Allium cepa*. Chem Nat Comp 22:553–556

Kuo PC, Lin TC, Yang CW, Lin CL, Chen GF, Huang JW (2010) Bioactive saponin from tea seed pomace with inhibitory effects against Rhizoctonia solani. J Agric Food Chem 58:8618–8622

Lanzotti V (2005) Bioactive saponins from *Allium* and *Aster* plants. Phytochem Rev 4:95–110

Lei Z, Watson BS, Huhman D, Yang DS, Sumner LW (2019) Large-scale profiling of saponins in different ecotypes of *Medicago truncatula*. Front Plant Sci 10:850

Li N, Li X, Feng Z (2008) Chemical constituents of tea seed saponins part from *Camellia sinensis* cultivated in China. J Shenyang Pharm Univ 25:544–548

Li N, Ma Z-J, Chu Y, Wang Y, Li X (2013) Phytochemical analysis of the triterpenoids with cytotoxicity and QR inducing properties from the total tea seed saponin of *Camellia sinensis*. Fitoterapia 84:321–325

Li X, Zhao J, Peng C, Chen Z, Liu Y, Xu Q, Khan IA, Yang S (2014) Cytotoxic triterpenoid glycosides from the roots of *Camellia oleifera*. Planta Med 80:590–598

Li X, Zhao J, Li X, LiuY XQ, Khan IA, Yang S (2015) New triterpenoid glycosides from the roots of *Camellia oleifera* Abel. Helv Chim Acta 98:496–508

Lu Y, Umeda T, Yagi A, Sakata K, Chaudhuri T, Ganguly DK, Sarma S (2000) Triterpenoid saponins from the roots of tea plant (*Camellia sinensis* var. assamica). Phytochemistry 53:941–946

Martyniuk S, Biały Z (2008) Effects of saponins from Medicago arabica on in vitro growth of *Cephalosporium gramineum*. Phytopathol Pol 49:49–55

Matsuda H, Nakamura S, Fujimoto K, Moriuchi R, Kimura Y, Ikoma N, Hata Y, Muraoka O, Yoshikawa M (2010) Medicinal Flowers. XXXI. Acylated oleanane-type triterpene saponins, Sasanquasaponins I-V, with antiallergic activity from the flower buds of *Camellia sasanqua*. Chem Pharm Bull 58:1617–1621

Matsuda H, Hamao M, Nakamura S, Kon'i H, Murata M, Yoshikawa M (2012) Medicinal flowers. XXXIII. Anti-hyperlipidemic and anti-hyperglycemic effects of chakasaponins I-III and structure of chakasaponin IV from flower buds of Chinese tea plant (*Camellia sinensis*). Chem Pharm Bull 60:674–680

Morikawa T, Li N, Nagatomo A, Matsuda H, Li X, Yoshikawa M (2006) Triterpene saponins with gastroprotective effects from tea seed (the seeds of *Camellia sinensis*). J Nat Prod 69:185–190

Morikawa T, Nakamura S, Kato Y, Muraoka O, Matsuda H, Yoshikawa M (2007) Bioactive saponins and glycosides. XXVIII. new triterpene saponins, foliatheasaponins I, II, III, IV, and V, from tencha (the leaves of *Camellia sinensis*). Chem Pharm Bull 55:293–298

Mostafa A et al (2013) Aginoside saponin, a potent antifungal compound, and secondary metabolite analyses from *Allium nigrum* L. Phytochem Lett 6:274–280

Mskhiladze L, Legault J, Lavoie S, Mshvildadze V, Kuchukhidze J, Elias R, Pichette A (2008) Cytotoxic steroidal saponins from the flowers of *Allium leucanthum*. Molecules 13:2925–2934

Murakami T, Nakamura J, Matsuda H, Yoshikawa M (1999) Bioactive saponins and glycosides. XV. Saponin constituents with gastroprotective effect from the seeds of tea plant, *Camellia sinensis* L. var. *assamica* Pierre, cultivated in Sri Lanka: structures of assamsaponins A, B, C, D, and E. Chem Pharm Bull 47:1759–1764

Murakami T, Nakamura J, Kageura T, Matsuda H, Yoshikawa M (2000) Bioactive saponins and glycosides. XVII. Inhibitory effect on gastric emptying and accelerating effect on gastrointestinal transit of tea saponins: structures of assamsaponins F, G, H, I, and J from the seeds and leaves of the tea plant. Chem Pharm Bull (Tokyo) 48:1720–1725

Myose M, Warashina T, Miyase T (2012) Triterpene saponins with hyaluronidase inhibitory activity from the seeds of *Camellia sinensis*. Chem Pharm Bull 60:612–623

Nakamura S, Moriura T, Park S, Fujimoto K, Matsumoto T, Ohta T, Matsuda M, Yoshikawa H (2012) Melanogenesis inhibitory and fibroblast proliferation accelerating effects of noroleanane- and oleanane-type triterpene oligoglycosides from the flower buds of *Camellia japonica* (1). J Nat Prod 75:1425–1430

Ohta T, Nakamura S, Nakashima S, Matsumoto T, Ogawa K, Fujimoto K, Fukaya M, Yoshikawa M, Matsuda H (2015) Acylated oleanane-type triterpene oligoglycosides from the flower buds of *Camellia sinensis* var. assamica. Tetrahedron 71:846–851

Peng J-P, Yao X-S, Tezuka Y et al (1996) Furostanol glycosides from bulbs of *Allium chinense*. Phytochemistry 41:283–285

Pérez AJ, Kowalczyk M, Simonet AM, Macias FA, Oleszek W, Stochmal A (2013) Isolation and structural determination of triterpenoid glycosides from the aerial parts of alsike clover (*Trifolium hybridum* L.). J Agric Food Chem 61:2631–2637

Pollier J, Morreel K, Geelen D, Goossens A (2011) Metabolite profiling of triterpene saponins in *Medicago truncatula* hairy roots by liquid chromatography Fourier transform ion cyclotron resonance mass spectrometry. J Nat Prod 74:1462–1476

Sagesaka YM, Uemura T, Watanabe N, Sakata K, Uzawa J (1994) A new glucuronide saponin from tea leaves (*Camellia sinensis* var. sinensis.). Biosci Biotechnol Biochem 58:2036–2040

Sang SM, Zou M, Xia Z et al (2001) New spirostanol saponins from Chinese chives (*Allium tuberosum*). J Agric Food Chem 49:4780–4783

Sashida Y, Kawashima K, Mimaki Y (1991) Novel polyhydroxylated steroidal saponins from *Allium giganteum*. Chem Pharm Bull 39:698–703

Sparg SG, Light ME, van Staden J (2004) Biological activities and distribution of plant saponins. J Ethnopharmacol 94:219–243

Sugimoto S, Chi G, Kato Y, Nakamura S, Matsuda H, Yoshikawa M (2009) Medicinal flowers. XXVI.1 structures of acylated oleanane-type triterpene oligoglycosides, yuchasaponins A, B, C, and D, from the flower buds of *Camellia oleifera*-gastroprotective, aldose reductase inhibitory, and radical scavenging effects. Chem Pharm Bull 57:269–275

Tava A, Pecetti L (2012) Chemical investigation of saponins from twelve annual *Medicago* species and their bioassay with the brine shrimp *Artemia salina*. Nat prod commun 7(7):837–840

Tava A, Mella M, Avato P, Argentieri MP, Bialy Z, Jurzysta M (2005) Triterpenoid glycosides from leaves of *Medicago arborea* L. J Agric Food Chem 53:9954–9965

Teshima Y, Ikeda T, Imada K, Sasaki K, El-Sayed MA, Shigyo M, Tanaka S, S-I/ I (2013) Identification and biological activity of antifungal saponins from shallot (*Allium cepa* L. Aggregatum Group). J Agric Food Chem (31):7440–7445

Trdá L, Janda M, Macková D, Pospíchalová R, Dobrev PI, Burketová L, Matušinsky P (2019) Dual Mode of the saponin aescin in plant protection: antifungal agent and plant defense elicitor. Front Plant Sci 10:1408

Uddin NM, Sharma G, Yang J-L, Choi HS, Lim SI, Kang KW, Oh WK (2014) Oleanane triterpenes as protein tyrosine phosphatase 1B (PTP1B) inhibitors from *Camellia japonica*. Phytochemistry 103:99–106

Varughese T, Manir MM, Rahaman M, Kim JK, Lee B-G, Moon S-S (2011) Tea triterpenoidal saponins from the roots of *Camellia sinensis* have inhibitory effects against alcohol dehydrogenase. Planta Med 77:2029–2036

Vincken J-P, Heng L, Groot A, Gruppen H (2007) Saponins, classification and occurrence in the plant kingdom. Phytochemistry 68:275–297

Vo NN, Fukushima EO, Muranaka T (2017) Structure and hemolytic activity relationships of triterpenoid saponins and sapogenins. J Nat Med 71(1)

Vollerner YS, Abdullaev ND, Gorovits MB et al (1984) Steroid saponins and sapogenins of *Allium* XX. Structure of karataviosides E and F. Chem Nat Comp 20:64–68

Vollerner YS, Kravets SD, Shaskov AS et al (1989) Steroids of the spirostan and furostan series from plants of the genus *Allium* XXVI. Structure of anzurogenin and anzuroside from the collective fruits of *Allium suvorovii* and *Allium stipitatum*. Chem Nat Comp 25:431–435

Wu J, Zhao J, Liu Y, Li X, Xu Q, Feng Y, Khan IA, Yang S (2015) Five new triterpenoid saponins from the roots of *Camellia oleifera* C. Abel with cytotoxic activities. Phytochem Lett 13:379–385

Wu JP, Kang NX, Zhang MY, Gao HW, Li XR, Liu YL et al (2018) Oleiferoside W from the roots of *Camellia oleifera* C. Abel, inducing cell cycle arrest and apoptosis in A549 cells. J Asian Nat Prod Res 20:793–806

Yang C-R, Zhang Y, Jacob MR, Khan S, Zhang YJ, Li X-C (2006) Antifungal activity of C 27 steroidal saponins. Antimicrob Agents Chemother 50:1710–1714

Yang WS, Ko J, Kim E, Kim JH, Park JG, Sung NY, Kim HG, Yang S, Rho HS, Hong YD, Shin SS, Cho JY (2014) 21-O-Angeloyltheasapogenol E3, a novel triterpenoid saponin from the seeds of tea plants, inhibits macrophage-mediated inflammatory responses in a NF-κB-dependent manner. Media Inflamm:1–9

Yang P, Li X, Liu YL, Xu QM, Li YQ, Yang SL (2015) Two triterpenoid glycosides from the roots of *Camellia oleifera* and their cytotoxic activity. J Asian Nat Prod Res 17:800–807

Yoshikawa M, Harada E, Murakami T, Matsuda H, Yamahara J, Murakami N (1994) Camelliasaponins B_1, B_2, C_1 and C_2, new type inhibitors of ethanol absorption in rats from the seeds of *Camellia japonica* L. Chem Pharm Bull 42:742–744

Yoshikawa M, Murakami T, Yoshizumi S, Murakami N, Yamahara J, Matsuda H (1996) Bioactive saponins and glycosides. V. acylated polyhydroxyolean-12-ene triterpene oligoglycosides, camelliasaponins A1, A2, B1, B2, C1 and C2, from the seeds of *Camellia japonica* L.: structures and inhibitory activity on alcohol absorption. Chem Pharm Bull 44:1899–1907

Yoshikawa M, Morikawa T, Li N, Nagatomo A, Li X, Matsuda H (2005) Bioactive saponins and glycosides. XXIII. Triterpene saponins with gastroprotective effect from the seeds of *Camellia sinensis*-theasaponins E3, E4, E5, E6, and E7. Chem Pharm Bull 53:1559–1564

Yoshikawa M, Morikawa T, Nakamura S, Li N, Li X, Matsuda H (2007) Bioactive saponins and glycosides. XXV. Acylated oleanane-type triterpene saponins from the seeds of tea plant (*Camellia sinensis*). Chem Pharm Bull 55:57–63

Yoshikawa M, Sugimoto S, Nakamura S, Matsuda H (2008) Medicinal flowers. XXII. Structures of chakasaponins V and VI, chakanoside I, and chakaflavonoside A from flower buds of Chinese tea plant (*Camellia sinensis*). Chem Pharm Bull 56:1297–1303

Yu X-L, He Y (2018) Optimization of tea-leaf saponins water extraction and relationships between their contents and tea (*Camellia sinensis*) tree varieties. Food Sci Nutr 6:1734–1740

Yu Z, Zhang T, Zhou F, Xiao X, Ding X, He H, Rang J, Quan M, Wang T, Zuo M, Xia L (2015) Anticancer activity of saponins from *Allium chinense* against the B16 melanoma and 4T1 breast carcinoma cell. Evid Based Complement Altern Med 2015:Article ID 725023

Yuan L, Ji TF, Li CJ et al (2009) Two new steroidal saponins from the seeds of *Allium cepa* L. J Asian Nat Prod Res 11:213–218

Zhang XF, Han Y-Y, Bao G-H, Ling T-J, Zhang L, Gao L-P, Xia T (2012) A new saponin from tea seed pomace (*Camellia oleifera* Abel) and its protective effect on PC12 cells. Molecules 17:11721–11728

Zhang H, Zhang W, Jiang J, Du J, Yang F (2015) Comparison of purification methods and component identification of tea saponin. China Oils Fats 40:94–98

Zhang ZY, Wu JP, Gao BB, Ren HT, Liu YL, Li XR et al (2016) Two new 28-nor-oleanane-type triterpene saponins from roots of *Camellia oleifera* and their cytotoxic activity. J Asian Nat Prod Res 18:669–676

Zhou H, Wang CZ, Ye JZ, Chen HX (2014) New triterpene saponins from the seed cake of *Camellia Oleifera* and their cytotoxic activity. Phytochem Lett 8:46–51

Zong JF, Wang R (2015) Novel triterpenoid saponins from residual seed cake of *Camellia oleifera* Abel. show anti-proliferative activity against tumor cells. Fitoterapia 107:7–13

Zong JF, Peng YR, Bao GH, Hou RY, Wan XC (2016) Two new oleanane-type saponins with anti-proliferative activity from *Camellia oleifera* Abel. seed cake. Molecules 21:188

Method of Estimation in Biological Sample 7

Abstract

Saponins are commonly found in adequate amounts in the root tissue of plant, however recent studies have reported that saponins can be also found in considerable amounts in plant aerial tissues such as leaf and stem. Thus, quantification of total saponin contents in different plant species and organs are very important to understand their biological functions in plant defense. There are several methods have been developed for measuring saponin contents in medicinal as well as crop plant species. The classical colorimetric and biological methods are remain popular methods for saponin quantification. However, biological and colorimetric determinations of saponin contents doesn't provide accurate information and sometimes might resulted in a misleading information, due the large structural variation of individual saponins not only within different species, but even also among same species. Thus, more sensitive methods have been recently introduced to measure and quantify saponin contents in different plant extracts. High performance (HP)-thin-layer chromatography (TLC) on normal (HPTLC) or reversed-phase (two-dimension, 2D-HPTLC) provides more precise and reliable saponin qualitative information, especially when these HPTLC methods are combined with a computer flying-spot scanner with dual-wavelength. After screening the saponin profile on the TLC, a 2D-analytical software can applied for the quantification of saponin level in plant extracts. However, for reliable measurements a proper saponin standards must be run with the saponin extracts for comparative analysis. Standardization and identification of the peaks by HPLC chromatograms has been also developed for saponin quantification, which relay on the comparisons of the retention times with those observed for authentic standards. On the other hand, there are limited applications of gas chromatography (GC) for quantification and determination of saponin compounds, due to the high molecular weights of the saponin compounds. In this chapter we will discussed some of these methods and the amount of saponin detected in different plant species.

M. Abdelrahman, S. Jogaiah, *Bioactive Molecules in Plant Defense*,
https://doi.org/10.1007/978-3-030-61149-1_7

7.1 Introduction

The initial determination of saponin compounds and other bioactive metabolites in plant materials was mainly based on the chemical or biological features (Van Atta et al. 1961; Naidu et al. 2011). For example, the foaming feature of the most of the saponin compounds has been used as specific marker assay for plant saponin content. However, some saponins with two or three sugar residues doesn't produced a steady foam; on the other hand, some plant extracts, which doesn't not have saponins can form froth when mixed with water solution, thus foaming test might provide misleading information. The haemolytic activity, an early biological marker test for the saponin, has been used as a semiquantitative assay for the determination of saponin in an extract (Mackie et al. 1977). In brief, saponin-rich extracts are being mixed with erythrocytes or blood in 0.9% NaCl solution inducted for 20–24 h, and centrifuged. Then the presence of the haemoglobin in the supernatant can be used as an indicator for saponin haemolysis activity. The haemolytic index (HI), which is defined by the European Pharmacopoeia as the total number of milliliters of blood that can be haemolysed by addition of 1 g of crude saponins. In general two saponin mixtures derived from Saponin white and *Gypsophila paniculata* L extracts are commonly used as standard reference with 15,000 and 30,000 HI, respectively. For easy measurement the HI of saponin haemolytic activity can be measured by using the following equation:

$$\mathrm{HI} = \mathrm{HIstd} \times \mathrm{a}/\mathrm{b}$$

where HIstd is the HI of reference saponin compound, and a and b are the lowest concentrations of examined saponin and reference saponin, respectively. Haemolytic activity of saponins can also be measured by using TLC plates (Wagner et al. 1986; Khalil and El-Adawy 1994).After development of saponin profile on the TLC, the TLC plants must be completely dried up from the solvents, and then a thin layer of gelatin-blood solution can be sprayed over the TLC plate in almost similar pattern. The TLC plates can be incubated for few hours, and the white spots on the TLC plates can be used as indicator for the saponin haemolytic activity. Although biological methods are simple and can be applied in different labs without need of sophisticated tools, these methods cannot distinguish between different saponin compounds (Oleszek 2002). On the other hand TLC, GC, HPLC, LC/NMR, LC/MS as well as capillary electrophoresis (CE) techniques have been recently used for the precise measurements of saponins in various plant extracts. Specifically the rapid initial screening of the crude saponin contents in various plant extracts can be screened by using LC/MS, LC/NMR and CE techniques, which can provide preliminary information on the nature and the content of saponin constituents in the extract (Oleszek 2002; Kawahara et al. 2016).

7.2 Determination of Saponins Using TLC

1D- or 2D-TLC methods are powerful techniques, which have been used effectively in the determination and separation of a considerable number of saponin compounds in different plant extracts. However, the main problem with 1D- or 2D-TLC techniques is the parallel running of an appropriate saponin standards and appropriate color spraying agents. Another problem with 1D- or 2D-TLC techniques is that the detection of saponin spots needs sophisticated instrumentation and software for data acquisition and handling to scan the saponin profile in the TLC plate at high speed. The later problem can be solved by using a computer flying-spot scanner with a dual-wavelength coupled with 2D-analytical software. For instance, AR-2000 radio-TLC-imaging scanner has been used for the finding of radiolabeled compounds in TLC plates. Also Tie-xin and Hong (2008) were able to develop an image analysis software. The developed system was validated by using quantitative assay of cichoric acid developed on polyamide TLC with $CHCl_3$-MeOH-CH_2O_2-H_2O (3:6:1:1, v:v:v) as the mobile phase, and 3% aqueous aluminum chloride solution as the visualizing reagent. The developed TLC images were captured under dark condition by a digital camera using UV lamp. The identified spot was then converted into corresponding peak area and the cichoric acid spot is integrated and used for quantitation. Based on this TLC methods a considerable number of saponin compounds have been detected (Table 7.1).

CHCl3:MeOH:H_2O or *n*-butanol: $C_2H_4O_2$:H_2O are the most common solvent system used for silica gel plates developments (Mostafa et al. 2013; Abdelrahman et al. 2014). On the other hand, Carr–Price and Liebermann–Burchard reagents,

Table 7.1 List of the saponin compounds identified by the mean of TLC-densitometry

Plates	Solvent system	Saponin compound	Reference
Silica gel	MeOH:H_2O (55:45)	Cucurbitacin B, D,E, I	Gorski et al. (1985)
Silica gel	EtOAc:C_6H_6 (75:25)	Cucurbitacin C	Gorski et al. (1986)
Silica gel G,H	$CHCl_3$:MeOH:H_2O (65:35:10)	Ginsenosides	Zhang et al. (1983)
Silica gel LS	BuOH:OHAc:H2O (4:1:5)	Gypsosides	Tagiev and Ismailov (1986)
Silica gel 60G	$CHCl_3$:MeOH:H_2O (65:35:10)	Soyasaponins A, B	Gurfinkel and Rao (2002)
Silica gel 60G	$CHCl_3$:MeOH:OHAc:H_2O	Asiaticoside Madecassoside	James and Dubery (2011)
Silica gel F254	$CHCl_3$:MeOH:H_2O (8:7:1)	Oleanane-derived sapogenols	Podolak et al. (2013)
Silica gel 60 W F254	$C_4H_8O_2$:MeOH:H_2O:OHAc (100:20:16:1)	Soyasaponnins	Shawky and Sallam (2017)
Silica gel G	$CHCl_3$:Et2O:MeOH (30:10:1)	Oleanolic acid	Zhang (1995)
Silica gel 60 F254	$C_4H_8O_2$: H_2O:CH_2O_2 (5:1:1)	Primulasaponins	Coran and Mulas (2012)

phosphotungstic and *p*-anisaldehyde 1% H_2SO_4 in OHAc acid are the major visualization sprayers (Coulson 1958; Abdelrahman et al. 2017). In general, HPLC method is sufficiently precise for quality control observation, and thus can be used for the quantitative assay of different extraction series. For example, Chaicharoenpong and Petsom (2009) analyzed saponins in ten different powdered tea seed meal samples by using HPTLC-silica gel plates and $C_4H_8O_2$:MeOH:H_2O (6:3:1.5) as mobile solvent system and detection wavelength at 214 nm. The tea saponin peak areas were measured at R_f 0.40, and the observed results indicated that the levels of tea saponins in the tested ten samples were ranged between 13.1 and 21.1% w/w. Recently, Shawky and Sallam (2017) were able to measure isoflavones and soyasaponins and in soybean (*Glycine max*) by-products by HPTLC method. The silica gel plat was developed using $C_4H_8O_2$:MeOH:H_2O:OHAc (100:20:16:1) solvent system. Then, UV-absorbance measurement at 265 nm using multi-wavelength scanner was carried to measure daidzin, genistin, and glycitin levels, whereas 650 nm Vis-absorbance wavelength was used for the detection of soyasaponins I and III. The LOD (μg mL^{-1}) of genistin, daidzin, glycitin, soyasaponin I and soyasaponin III were 0.0318, 0.0502, 0.0449, 0.1143 and 0.096, respectively. Similarly, a simple technique for the measurement of saponin levels in legume plants, by TLC-densitometry was tested (Gurfinkel and Rao 2002). Initially, saponin profile was developed on a TLC plate, in the same time a reference soyasaponin was also run for comparative analysis (Gurfinkel and Rao 2002). The plate was treated with sulfuric acid and heated and violet spots density was correlated with the amount of saponin content. The saponin contents of dried navy beans (defatted soy flour and dried kidney beans were 0.32, 0.58, and 0.29%, respectively (Gurfinkel and Rao 2002). Similarly, HPTLC technique was applied for the measurement of primulasaponin I and II in various extracts (Coran and Mulas 2012). The HPTLC results showed that primulasaponin contents were ranged between 150 and 450 ng and the relative standard deviation (RSD) of repeatability and intermediate accuracy ranged between 0.8 and 1.4% (Coran and Mulas 2012).

7.3 Quantification of Saponins by HPLC

The normal- and reverse-phase (RP) HPLC are the most powerful and frequently used techniques for the separation, purification and identification of saponin compounds (Negi et al. 2011). However, RP-HPLC with C_{18} column is more preferred for saponin separation and quantification (Negi et al. 2011). In addition, NH_2 and carbohydrate-modified columns have been also shown to be relatively successful in the identification of few steroidal saponins, however they have shown very effective separation of glycoalkaloids (Xu and Lin 1985; Saito et al. 1990). The solanine and chaconines also being well separated with Bondapack NH_2 column using RP-HPLC mode (Bushway et al. 1979). In addition, the resolution of closely related saponins can be enhanced by using hydroxyapatite, which is more hydrophilic compared with silica gel, and thus can allow the separation of glycosides with high similar structures (Kasai et al. 1987). In the following Table 7.2 we are

Table 7.2 List of saponin compounds identified by HPLC

Compound	Mobile system	Reference
Avenacosides	MeCN-H_2O	Kesselmeier and Strack (1981)
Diosgegnin	Hexane-iso-PrOH	Tal et al. (1984)
Saikosaponin-a and chikusetsusaponin V	MeOH–H_2O–$C_2H_4O_2$	Kimata et al. (1985)
Cucurbitacins	MeOH–H_2O	Gorski et al. (1986)
Madecassoside, Asiaticoside, Hederacoside C, Chrysantellin A, β-Escin, Echinocystic acid-3-glucoside and Gypsogenin-3-glucuronide	Acetonitrile-H_2O	Burnouf-Radosevich and Delfel (1986)
Soyasaponins I, II, III and IV	Acetonitrile-H_2O with 0.025% trifluoroacetic acid	Berhow et al. (2002)
Soyasaponins	H_2O:CH_2O_2 and MeOH:CH_2O_2	Gu et al. (2002)
Escin Ia, isoescin Ia, escin Ib, and isoescin Ib	MeOH-H_2O-$C_2H_4O_2$	Wei et al. (2005)
Soyasaponins	Acetonitrile-water	Lin and Wang (2006)
Ginsenosides, asiaticoside, deglucoruscoside and ruscoponticoside C	Acetonitrile-H_2O-MeOH with 1% acetic acid	Kite et al. (2007)
Azukisaponin I, II, III, IV, V, and VI	Acetonitrile-H_2O	Liu et al. (2017)
Saikosaponin	Acetonitrile-H_2O	Liu et al. (2018)
Sopogenin	Acetonitrile-H_2O	Soni et al. (2020)

summarizing some of the saponin compounds identified by HPLC method and the mobile phase being applied.

Although HPLC is very powerful technique for saponin identification, the main technical problem in HPLC method is the UV detection range. For example, saponin compounds such as cucurbitacins and glycyrrhetinic acid and its glycosides have maximum absorption within the UV range, and thus can be easily detected at 254 nm. However, most of the saponin compounds doesn't have chromophores that are needed for UV measurement, and thus, both detection and separation of saponin glycosides as well as their aglycones have to be measured at lower UV ranging from 200 to 210 nm (Nyakudya et al. 2014). On the other hand, the lower UV detection range, decreases the efficiency of selection of difference solvent and gradient systems, and thus, at lower wavelengths acetonitrile has lower absorption therefore its mostly preferred than MeOH as mobile phase (Nyakudya et al. 2014).

To overcome the detection problems at lower wavelength, pre-column derivatization of saponins can be an efficient way to attach a chromophore to saponin compounds and facilitates UV detection at higher wavelength 254 nm (Peng et al. 2008). For example, sarsasapogenin extracted from *Rhizoma Anemarrhenae*, saponin was pre-derivatization with benzoyl chloride derivate. Then the chromatography was conducted on Agilent HPLC C_{18} column, using methanol-water as the mobile phase with 230 nm detection wavelength (Peng et al. 2008). The sarsasapogenin content ranged between 0.20 and 4.00 g L^{-1}. (Peng et al. 2008). Similarly, gas chromatography mass spectrometry (GC-MS) system was used to measure the level of standard mixture of eight triterpene saponins (cholesterol, α-amyrin, β-amyrin, oleanolic acid, lupeol, betulinic acid, hederagenin, and α-epoxi-β-amyrin) polyphenols, amino acids, carbohydrates as well as blue berry extracts were derivative by trimethylsilyl cyanide (TMSCN) and silylation reagent N-methyl-N-(trimethylsilyl) trifluoroacetamide (MSTFA) before the injection (Khakimov et al. 2013). The TMSCN derivatization was 54 times more sensitive than MSTFA derivatization (Khakimov et al. 2013). The medicagenic acid glycosides, bidesmosidic and monodesmosidic forms can be chromatographed on a C_{18} column after derivatization (Nowacka and Oleszek 1997). Zanhic acid tridesmoside in the alfalfa had both –COOH glycosylated, and due to the high polarity of this Zanhic acid tridesmoside, the determination at 210 was also very difficult (Nowacka and Oleszek 1997). To overcome this problem, the hydrolysis of zanhic acid tridesmoside using alkaline system can be conducted before the derivatization with 4-bromophenacyl bromide, and then the obtained prosapogenin derivatives can be chromatographed at wavelength of 260 nm (Nowacka and Oleszek 1997). Likewise, successful results have been also achieved by derivatization of saponin compounds using 4-bromophenacyl bromide in the presence of cyclic ether. For example, *Phytolacca dodecandra* oleane saponins were derivative with 4-bromophenacyl bromide (Slacanin et al. 1988), however saponin compound must possess at least one carboxyl group, either at the sugar part or aglycone. Thereby, this method only suitable for the detection of few groups of saponin compounds (Oleszek 2002). Although pre-derivatization steps can significantly improve the detection process, it also might create some technical problems due to the differentiated rate of substitution of functional –OH groups and steric shape of the saponin molecule, and thus the derivative saponin complex mixtures generate many peaks on the chromatogram and their quantification and interpretation are not straight forward (Oleszek 2002).

Although HPLC methods have been widely used for saponin detection and quantification, the HPLC-UV methods don't usually guarantee the identification of individual peak because the detection UV spectral wavelength are not always specific (Oleszek 2002; Khakimov et al. 2013). Thus, identification of unknown saponin compounds depend on the specific retention times of unknown peaks with the retention times of related saponin standards. In order to improve the efficiency of saponin detection, hyphenated, methods through HPLC combined with various spectroscopic detection methods have been recently developed (Khakimov et al. 2013). Several studies used hyphenated HPLC analytical platforms, including GC-MS, LC-MS/MS, LC-nuclear manganic resonance (NMR)/MS and electrospray (LC-ES-MS) techniques to screen saponins in different plant extracts (Muir et al. 2000). For example, comparative saponin and saponin aglycone profiles isolated from hydrolyzed samples of two different cultivars of *Barbarea vulgaris* plants 'glabrous' and 'pubescent' which are known to be different in their insect resistance ability, using GC-MS, LC-MS/MS, and LC-SPE-NMR/MS methods, was conducted. The obtained results showed significant differences between insect-resistant and susceptible cultivars in terms of total saponin contents, and also in the types of aglycones as well as the numbers of sugar residues (Khakimov et al. 2013). Identification of two previously known insect-deterrent saponins, oleanolic acid and hederagenin cellobioside indicated that the the LC-MS/MS and LC-SPE-NMR/MS methods are reliable and provide rapid screening of bioactive compounds in different plant extracts (Khakimov et al. 2013). Likewise, HPLC coupled with diode array detection (DAD), and ion trap electrospray mass spectrometry (HPLC-DAD-ESI-MS/MS) was implemented for the analysis of triterpene saponins in *Zornia brasiliensis*. The retention times, and high-resolution MS determination and fragmentation, revealed 35 oleanane-triterpene saponin compounds were initially identified in *Z. brasiliensis* (Nascimento et al. 2019). To quantify, the total saponin contents, and types of individual saponins found in leaf extracts of *Cassytha filiformis* GC-MS method was applied (Edewor et al. 2016). First, the leaf tissues were extracted with *n*-hexane and MeOH, then the MeOH extract was partitioned using *n*-butanol-H_2O system. Then *n*-butanol fraction rich with saponin was examined by UV spectrometry using ginsenoside as the authentic standard at 550 nm detection range (Edewor et al. 2016). The saponin-enriched butanol fraction was also subjected to GC-MS and the total saponin contents of the MeOH extract was 73.47 μg ginsenoside Rb1 equivalent/g extract. In addition, cholestan-7-one and cyclic 1,2-ethanedienyl acetal were the most abundant saponin compounds have been identified in the *n*-butanol fraction (Edewor et al. 2016). Likewise, saponins, terpenoids, flavonoids, and alkaloids in the extract of *Vernonia cinerea* leaves were analyzed by LC-Q-TOF-MS analysis. The results showed that 221 compounds were initially assigned in the extract of *V. cinerea* leaves, including 13 saponin compounds, 108 terpenoids, 64 flavonoids and 36 alkaloids.

In general, the extraction process and saponin standards are very important steps for the precise quantification of the saponin compounds in complex extract and in Table 7.3 we summarized some of saponin standard price list, and also in Fig. 7.1 we develop a schematic graph of the saponin extraction process.

Table 7.3 List of saponin standards and their average price per mg

Saponin name	HPLC purity (%)	Price \$ mg^{-1}	Company
Ginsenoside-Ro	≥98	4.9	Star-Ocean
Ginsenoside-Rb1	≥98	0.95	Star-Ocean
Ginsenoside-Rb2	≥98	3.7	Star-Ocean
Ginsenoside-Rb3	≥98	3.7	Star-Ocean
Ginsenoside-Rc	≥98	3.7	Star-Ocean
Ginsenoside-Rd	≥98	2.8	Star-Ocean
Ginsenoside-Re	≥98	0.95	Star-Ocean
Diosgenin	≥93	2.5	Merck
Soyasaponin I	≥98	12.0	Merck
Soyasapogenol B	≥98	11.0	Merck
Soyasapogenol A	≥98	11.5	Merck
Ginsenoside-Rg3	≥98	4.5	Merck

Note: the above price can varied according to company, and the extract price update can be found in the company homepages

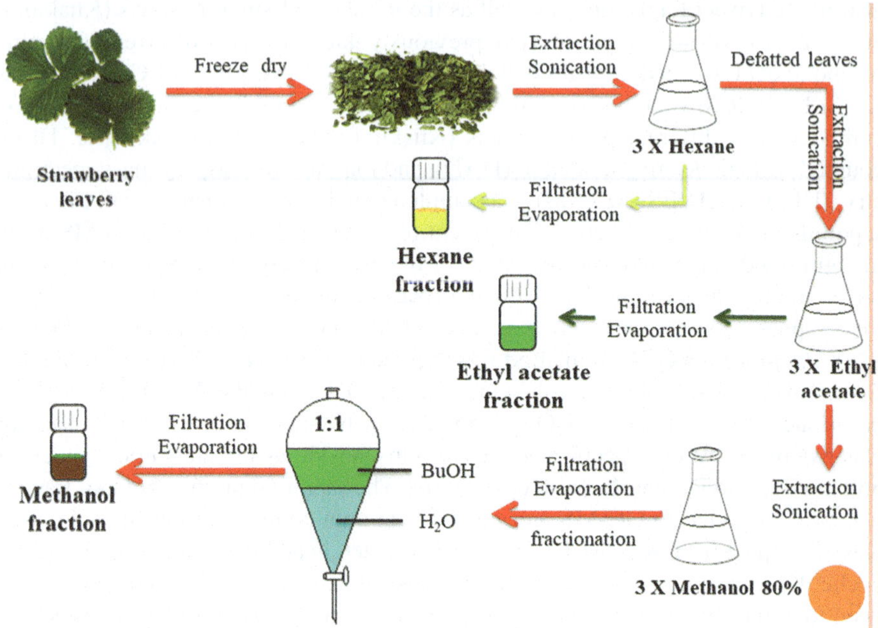

Fig. 7.1 Schematic model for the saponin extraction from strawberry leaves

7.3.1 Determination of Saponins in Yucca (*Yucca schidigera*) Extract

A spectrophotometric technique was established for the quantification of total crude saponin contents in Yucca extract (Uematsu et al. 2000). First, saponin fraction was isolated from Yucca extract using column chromatography. The saponin fraction was hydrolyzed with 2 mol L^{-1} of HCL:EtOH mixture (1:1, v:v) to obtain the sapogenin aglycon. Then, the sapogenin contents based on the color reactions with acidic-anisaldehyde reagent were measured at 430 nm using spectrophotometer (Uematsu et al. 2000). The detailed method of extraction and quantification can be described in details as follow:

7.3.1.1 Sample Preparation

Twenty mL of HP-20 resin was mixed with in MeOH for 24. Then the HP-20 was packed with MeOH in a glass column with 15 mm diameter. After filling the column, the resin washed with 100 mL MeOH and 200 mL H_2O successively. Yucca crude extract was dissolved in small volume of H_2O and loaded into the column. The crude extract has been washed with 100 mL H_2O and 100 mL 40% MeOH, successively as mobile solvent system. In order to elute saponin fraction, the column run with 100 mL 95% MeOH solution. The obtained saponin faction was further concentrated by removing the MeOH solvent using rotary evaporator under reduced pressure. The saponin residue was then dissolved with 20 ml methanol and used for acid hydrolysis.

7.3.1.2 Acid Hydrolysis

Twenty mL of saponin crude extract was placed in flask and MeOH solvent was removed using rotary evaporator under reduced pressure. A mixture of EtOH:HCL 20 mL (1:1) was added to the saponin residue, and kept for acid hydrolysis up to 3 h at 90°C. After the solution cooled down, 80 mL diethyl ether solvent was added. The saponin fraction in the diethyl ether layer was collected and washed with 20 mL H_2O. The organic layer was dried over anhydrous sodium sulfate and the ether solvent was removed using rotatory evaporator under reduced pressure. The sapogenin in the final residue was dissolved with 10 mL ethyl acetate for spectrophotometric analysis.

7.3.1.3 Spectrophotometric Determination

To measure the sapogenin contents, a color regent consisted of 0.5 mL *p*-anisaldehyde, 99.5 mL ethyl acetate and 50 mL concentrated sulfuric acid were prepared carefully. Sapongenin solution was diluted initially with ethyl acetate to obtain 2.5–10 mg mL^{-1} sapogenins. 2 mL of the diluted sapogenin solution was transferred into 10 mL test tube and 1 mL coloring reagent was added and the test tube sealed with a glass stopper. The glass test tube was incubated in a water bath at 60 °C for 10 min until a brownish-pink color developed. The absorbance of the solution was measured using ethyl acetate as blank. Total saponin contents was estimated using sapogenin (2–40 μg) standard curves (Uematsu et al. 2000). The hydrolyzed Yucca extract contained 2.7–3.0% sapogenin. The saponin value was calculated to be 5.6–6.4%, according to the molecular mass of sarsasapogenin and saponin YE-2 (Uematsu et al. 2000).

7.3.2 Determination of Saponin in *Camellia sinensis* and Genus *Ilex* Using HPLC

7.3.2.1 Saponin Extraction

In this method total saponin contents in tea leaves were measured using HPLC according to Kim and Wampler (2009). Briefly, five dried tea leaves, including black tea, green tea, Yaupon holly-containing caffeine, Yaupon holly caffeine free, and Yerba mate were grinded into fine powder using pestle and mortar. The fine powder samples were extracted by hot water for 10 min at 90°C. The tea crude extract was separated into three groups containing 20 mL of the infusion. The extract was treated with 3 mL of chloridric acid to yield an acid concentration of 4 mol L^{-1} prior to hydrolysis for 2 h. The saponin rich fraction was then isolated with same volume of $CHCl_3$ using separation funnel. The extraction steps were repeated three times, and the pooled saponin fractions were dried under reduced pressure.

7.3.2.2 HPLC Determination

The saponin fractions derived from the tea extracts were diluted threefold using deionized water. The diluted saponin was passed through a 0.45 μm PTFE filter before injection into the HPLC system. Agilent 1200 HPLC system was used to separate the polyphenolic compounds using a UV-Vis detector with and Acclaim 120-C_{18} column with a flow rate of 0.8 mL min^{-1}. O-phosphoric acid was used to adjust the pH of the mobile phase (100% H_2O) to 2.4, and it was run for 30 min at 0.8 mL min^{-1}. The contents of saponins in each tea infusions was measured at 280 nm absorbance using ursolic acid as external standard (Sigma Chemical Co., St. Louis, MO). The obtained results indicated that saponin contents were highest in yaupon holly followed by yerba mate, black tea, and green tea ranging 75.05–92.62 mg L^{-1} (Kim and Wampler 2009).

7.3.3 Determination of Saponin in Ophiopogon Japonicas Using HPLC

The levels of three saponin compounds, namely ophiopogonins B, D, and D', in the fibrous roots and tubers of *Ophiopogon japonicus* grown at two different geographical locations were quantified using HPLC-evaporative light scattering detector (ELSD) according to Li et al. (2016).

7.3.3.1 Sample Preparation

The tuber and fibrous root materials were heated and washed for 15 min at 105 °C to inhibit the enzymatic activity. Then, the tuber and fibrous root samples were dried at 60 °C, grounded using mixer and the obtained dry fine power was filtered through a 40-mesh-size filter, and stored in a dissector (Li et al. 2016).

7.3.3.2 Saponin Extraction

Two grams of the dry fine powder were extracted with 100 mL MeOH for 1 h, and filtered using cheesecloth. Then, the crude extract was concentrated by evaporating the MeOH solvent using a rotary evaporator at 65 °C under the reduced pressure. The cure residue was dissolved in 40 mL of deionized water and mixed with 20 mL of petroleum ether to remove oils and fats. The petroleum ether layer was removed, while water fraction was collected and then partitioned three times with 90 mL of H_2O-*n*-butanol (1:1, v:v). Water layer discarded and *n*-butanol fraction was collected and concentrated at 65 °C under reduced pressure. The final crude saponin extract was dissolved in 2 mL MeOH for the HPLC analysis (Li et al. 2016).

7.3.3.3 Determination of Steroidal Saponins

Mixed stock standard solution of ophiopogonins B, D, and D′ (0.082 mg, 0.111 mg, and 0.095 mg) has been prepared in 1 mL MeOH. After this a total of 2, 4, 6, 8, 10, 15, and 20 μL of the standard mixtures were diluted in MeOH for HPLC-ELSD analysis. Saponin calibration curves were developed by plotting the HPLC-ELSD peak areas compared with the respective concentration of each standard. To identify the saponin contents in the unknown samples, 10 μL of the extracts and saponin standards were injected into Waters 600E system equipped with an Agela Venusil ASB C_{18} column and Waters 2424 evaporative light-scattering detector. Acetonitrile (A) and water (B) solvent system at 1 mL min^{-1} flow rate was applied. Gradient system with 45% A to 55% A for 30 min, 55% A to 45% A for 5 min, and 45% A held for another 5 min was carried out. Ophiopogonin B and D' contents in the crude extracts of Hang Maidong (HMD) tubers were greater than their levels in the radix of Chuan Maidong (Li et al. 2016). On the other hand, ophiopogonin D saponin in Hang Maidong was two times lower than its respective content in the Chuan Maidong (Li et al. 2016).

7.3.4 Total Saponins in *Ilex paraguariensis* Extract

Gnoatto et al. (2005) describes an HPLC protocol for quantification of total saponin contents in *I. paraguariensis* aqueous extracts, using ursolic acid as external standard.

7.3.4.1 Sample Preparation and Saponin Extraction

Leaves of *I. paraguariensis* were collected and dried. 15 grams of the dried ground leaves were extracted for 10 min using hot distilled water. The aqueous extraction was then filtered through Whitman filter paper and the fill up to 100 mL with distilled water. To hydrolyze saponins, 100 mL of the aqueous solution were mixed with 15 mL chloridric acid, refluxed for 2 h and the sapogenin-rich fractions were extracted with 50 mL $CHCl_3$. The $CHCl_3$ extraction step was repeated four times. The pooled $CHCl_3$ fractions were concentrated using rotary evaporator under reduced pressure to obtain the sapogenin residue. The sapogenin residue was dissolved in 50 mL acetonitrile and kept for HPLC analysis.

7.3.4.2 HPLC Analysis

The sapogenin acetonitrile solution, was then diluted tenfold (1:9) with acetonitrile. The diluted solution was filtered through a 0.45 μm filter membrane and analyzed by HPLC. The total sapogenin contents were quantified using the calibration curves obtained by HPLC analysis of the standard solution of ursolic acid detected at 203 nm. The total saponins contents in *I. paraguariensis* extracts were of 352 μg mL^{-1} (Gnoatto et al. 2005).

7.3.5 Isolation and Characterization of Agenosoide Saponin from *Allium nigrum*

Spirostane-type glycoside namely aginoside was isolated from the root extracts of *A. nigrum* using 2D NMR, FABMS, HR-ESI-MS (Mostafa et al. 2013). The identified structure of the aginoside was 25(R,S)-5a-spirostan-2a,3b,6b-triol-3-O-b-D-glucopyranosyl-(1!2)-O-[b-D-xylopyranosyl-(1!3)]-O-b-D-glucopyranosyl-(1! 4)-b-D-galactopyranoside. In addition, the highest content of aginoside was 2.9 mg g^{-1} DW in the root tissue. In addition the antifungal activity of aginoside was evaluated showing strong inhibition to *Fusarium* and other phytopathogens. Spore germination assay, with different aginoside concentrations, showed also strong inhibition of spore germination in all tested pathogens, including *Botrytis cinerea* and *C. gloeosporioides*.

7.4 Conclusion

Chromatographic separation and quantification of saponin compounds in complex plant extracts are still a challenge. Until now, there is no single method that can be applied as general procedure for analysis of complex saponin mixtures. The HPLC combined with mass spectrometry is gaining much ground in saponin profiling, however the extraction process and proper derivatization need to be considered. Thus, a combination of several techniques is required to obtain single standard compounds.

References

Abdelrahman M, Hirata S, Ito SI, Yamauchi N, Shigyo M (2014) Compartmentation and localization of bioactive metabolites in different organs of *Allium roylei*. Biosci Biotechnol Biochem 78:1112–1122

Abdelrahman M et al (2017) RNA-sequencing-based transcriptome and biochemical analyses of steroidal saponin pathway in a complete set of *Allium fistulosum*—*A. cepa* monosomic addition lines. PLoS One 12:e0181784

Berhow MA, Cantrell CL, Duval SM, Dobbins TA, Mavnes J, Vaughn SF (2002) Analysis and quantitative determination of group B saponins in processed soybean product. Phytochem Anal 13:343–348

Burnouf-Radosevich M, Delfel NE (1986) High-performance liquid chromatography of triterpene saponins. J Chromatogr 368:433–438

Bushway RJ, Barden ES, Bushway AW, Bushway AA (1979) High-performance liquid chromatographic separation of potato glycoalkaloids. J Chromatogr A 178:533–541

Chaicharoenpong C, Petsom A (2009) Quantitative thin layer chromatographic analysis of the saponins in tea seed meal. Phytochem Anal 20:253–255

Coran SA, Mulas S (2012) Validated determination of primulasaponins in primula root by a high-performance-thin-layer-chromatography densitometric approach. J Pharm Biomed Anal 70:647–645

Coulson CB (1958) Saponins. I.-Triterpenoid saponins from lucerne and other species. J Sci Food Agric 9:281–288

Edewor TI, Owa SO, Ologan AO, Akinfemi F (2016) Quantitative determination of the saponin content and GC-MS study of the medicinal plant *Cassytha fiiformis* (linn.) leaves. J Coastal Life Med 4:154–156

Gnoatto SCB, Schenkel EP, Bassani VL (2005) HPLC method to assay total saponins in *Ilex paraguariensis* aqueous extract. J Braz Chem Soc 16:1678–4790

Gorski PM, Jaworski A, Shannon S, Robinson RW (1985) Rapid TLC and KPLC test for cucurbitacins. Genet Coop Rep 8:69–70

Gorski PM, Jaworski A, Shannon S, Robinson RW (1986) Rapid TLC and HPLC quantification of cucurbitacin C in cucumber cotyledons. HortScience 21:1034–1036

Gu L, Tao G, Gu W, Prior RL (2002) Determination of soyasaponins in soy with LC-MS following structural unification by partial alkaline degradation. J Agric Food Chem 50:6951–6959

Gurfinkel DM, Rao AV (2002) Determination of saponins in legumes by direct densitometry. J Agric Food Chem 50:426–430

James J, Dubery I (2011) Identification and quantification of triterpenoid centelloids in *Centella asiatica* (L.) urban by densitometric TLC. J Planar Chromatogr Modern TLC 24:82–87

Kasai R, Yamaguchi H, Tanaka O (1987) High-performance liquid chromatography of glycosides on a new type of hydroxyapatite column. J Chromatogr A 407:205–210

Kawahara Y, Hoshino T, Morimoto H, Shinizu T, Narukawa Y, Fuchino H, Kawahara N, Kiuchi F (2016) LC-MS-based quantification method for Achyranthes root saponins. J Nat Med 70:102–106

Kesselmeier J, Strack D (1981) High performance liquid chromatography analysis of steroidal saponins from *Avena sativa* L. Zeitschrift fur Naturforschung Sect C J Biosci 36:1072–1074

Khakimov B, Motawia MS, Bak S, Engelsen SB (2013) The use of trimethylsilyl cyanide derivatization for robust and broad-spectrum high-throughput gas chromatography-mass spectrometry based metabolomics. Anal Bioanal Chem 405:9193–9205

Khalil AH, El-Adawy TA (1994) Isolation, identification and toxicity of saponin from different legumes. Food Chem 50:197–120

Kim Y, Wampler DJ (2009) Determination of saponin and various chemical compounds in *Camellia sinensis* and genus *Ilex*. Sensus Tech Note (SEN-TN-0027)

Kimata H, Sumida N, Matsufuji N, Morita T, Ito K, Yata N, Tanaka O (1985) Interaction of saponin of bupleuri radix with ginseng saponin: solubilization of saikosaponin-a with chikusetsusaponin V (= ginsenoside-Ro). Chem Pharm Bull (Tokyo) 33:2849–2853

Kite GC, Porter EA, Simmonds MS (2007) Chromatographic behaviour of steroidal saponins studied by high-performance liquid chromatography-mass spectrometry. J. Chromatogr A 1148:177–183

Li XE, Wang YX, Sun P, Liao DO (2016) Determination of saponin content in Hang Maidong and Chuan Maidong via HPLC-ELSD analysis. J Anal Meth Chem 2016:7214607

Lin J, Wang C (2006) An analytical method for soy saponins by HPLC/ELSD. Food Sci 69:C456–C462

Liu R, Cai Z, Xu B (2017) Characterization and quantification of flavonoids and saponins in adzuki bean (*Vigna angularis* L.) by HPLC–DAD–ESI–MS analysis. Chem Central J 11:93

Liu Y-M, Zhou A, Yu N-J, Han R-C, Zhang W, Zhu Y-J, Cao Y, Li X-Y, Peng D-Y (2018) Simultaneous determination of five saponins in Bupleuri Radix by HPLC-DAD dual wavelength method. Zhongguo Zhong Yao Za Zhi 43:363–368

Mackie AM, Singh HT, Owen JM (1977) Studies on the distribution, biosynthesis and function of steroidal saponins in echinoderms. Comp Biochem Physiol Part B Comp Biochem 56:9–14

Mostafa A et al (2013) Aginoside saponin, a potent antifungal compound, and secondary metabolite analyses from *Allium nigrum* L. Phytochem Lett 6:274–280

Muir AD, Ballantyne KD, Hall TW (2000) LC-MS and LC-MS/MS analysis of saponins and sapogenins—comparison of ionization techniques and their usefulness in compound identification. In: Oleszek W, Marston A (eds) Saponins in food, feedstuffs and medicinal plants Proceedings of the Phytochemical Society of Europe, vol 45. Springer, Dordrecht

Naidu MM, Shyamala BN, Naik JP, Sulochanamma G, Srinivas P (2011) Chemical composition and antioxidant activity of the husk and endosperm of fenugreek seeds. LWT Food Sci Technol 44:451–456

Nascimento YM, Abreu LS, Lima RL, Costa VCO, de Melo JIM, Braz-Filho R, Silva MS, Tavares JF (2019) Rapid characterization of triterpene saponins from *Zornia brasiliensis* by HPLC-ESI-MS/MS. Molecules 24:2519

Negi JS, Singh P, Pant GJN, Rawat MSM (2011) High-performance liquid chromatography analysis of plant saponins: An update 2005-2010. Pharmacogn Rev 5:155–158

Nowacka J, Oleszek W (1997) High performance liquid chromatography of zanhic acid glycoside in alfalfa (*Medicago sativa*). Phytochem Analysis 3:227–230

Nyakudya E, Jeong JH, Lee NK, Jeong Y-S (2014) Platycosides from the roots of *Platycodon grandiflorum* and their health benefits. Prev Nutr Food Sci 19:59–68

Oleszek WA (2002) Chromatographic determination of plant saponins. J Chromatogr A 967:147–162

Peng Y, Luo F, Wang S, Li L, Sun Y, Pan W (2008) Determination of sarsasapogenin in Rhizoma Anemarrhenae with precolumn derivatization by HPLC. J Shenyng Pharm Univ 25:372–375
Podolak I, Hubicka U, Zuromska-Witek B, Janeczko Z, Krzek J (2013) Quantification of saponins in different plant parts of *Lysimachia* L. species by validated HPTLC-densitometric method. J Planar Chromatogr Modern TLC 26(3)
Saito K, Horie M, Hoshino Y, Nose N, Nakazawa H (1990) High-performance liquid chromatographic determination of glycoalkaloids in potato products. J Chromatogr 508:141–147
Shawky E, Sallam SM (2017) Simultaneous determination of soyasaponins and isoflavones in soy (*Glycine max* L.) products by HPTLC-densitometry-multiple detection. J Chrom Sci 55:1059–1065
Slacanin I, Marston A, Hostettmann K (1988) High-performance liquid chromatographic determination of molluscicidal saponins from *Phytolacca dodecandra* (Phytolaccaceae). J Chromatogr 448:265–274
Soni N, Singh VK, Singh DK (2020) HPLC characterization of molluscicidal component of *Tamarindus indica* and its mode of action on nervous tissue of *Lymnaea acuminate*. J. Ayurveda Integr Med 11:131–139
Tagiev SA, Ismailov AI (1986) Quantitative determination of gyposoide in roots of *Gypsophilla bicolor* Grossh. Rastit Resur 22:262–265
Tal DM, Patrick PH, Elliott W (1984) Bile acids. Lxx. preparative separation of kryptogenin from companion sapogenins by high performance liquid chromatography. J Liquid Chromatogr 7:2591–2603
Tie-xin T, Hong W (2008) An image analysis system for thin-layer chromatography quantification and its validation. J Chromatogr Sci 46:560–564
Uematsu Y, Hirata K, Saito K (2000) Spectrophotometric determination of saponin in Yucca extract used as food additive. J AOAC Int 83(6)
Van Atta GR, Guggolz J, Thompson CR (1961) Plant analysis, determination of saponins in alfalfa. J Agric Food Chem 9:77–79
Wagner H, Bladt S, Zgainski EM (1986) Plant Drug Analysis-a thin layer chromatography atlas. Springer, Berlin
Wei F, Ma L-Y, Cheng X-L, Lin R-C, Jin W-T, Khan IA, Lu JQ (2005) Preparative HPLC for Purification of Four Isomeric Bioactive Saponins from the Seeds of *Aesculus chinensis*. J Liquid Chromatogr Related Technol 28(5)
Xu C-J, Lin J-T (1985) Comparison of silica-, C18-, and NH2-Hplc columns for the separation of neutral steroid saponins from dioscorea plants. J Liquid Chromatogr 8:361–368
Zhang Q (1995) Determination of oleanolic acid in roots of *Achyranthes bidentata* by TLC-scan. Chin Pharm J 30:592–594
Zhang GD, Zhou ZH, Liu HY (1983) Analysis of ginseng. III. Isolation and determination of ginseng saponins. Acta Pharmaceutica Sinica 18:607–611

Genetic Engineering of Saponin Target Genes to Improve Yields

8

Abstract

Plant is a main source of natural products with diverse chemical structures and biological activities. Among different plant-derived natural products, saponins secondary metabolites are widely distributed across diverse plant species and have great potentials for the pharmaceutical industry, detergents, pesticides and plant disease management. Triterpene glycosides characterized by a 30 carbon atoms, namely oxidosqualene, a major precursor for triterpene aglycone (sapogenin), to which sugar chains are attached to yield the corresponding saponin compounds. However, saponin applications are often limited due to the low yields or accumulation in planta, inadequate of natural resources and the continuous need of the new compounds with superior biological activities. In addition, the biosynthesis and regulatory pathways of the saponin compounds in different plant species are still very limited. Thus development of new methods to improve and diversify the production of saponin glycosides, with a foresight into metabolic engineering, can be an alternative solution to avoid the problem associated with saponin large scale production. In this chapter, we will summarized and discussed the available information regarding saponin engineering in plant and their potential applications for plant disease resistance.

8.1 Biosynthesis of Plant Triterpene and Steroidal Saponins

In addition, to their role in plant disease resistance, saponins are important class of natural products for drug research due to their valuable pharmacological properties (Augustin et al. 2011). Therefore, much efforts have been made to understand the modes of action of saponin compounds, and their biological activities. However, lack of knowledge regarding biosynthetic genes involved in saponin biosynthesis in plants, is the main cause that hampered the investigation of saponins for bioengineering crop plants with enhanced disease resistances, as well as for industrial

M. Abdelrahman, S. Jogaiah, *Bioactive Molecules in Plant Defense*,
https://doi.org/10.1007/978-3-030-61149-1_8

production from plants natural resources (Augustin et al. 2011). In plants, all terpene-related molecules are mainly originated from the C5 precursor isopentenyl diphosphate (IPP) and dimethylallyl diphosphate (DMAPP). Both DMAPP and IPP precursors are originated from two distinct biosynthetic pathways, namely methylerythritol phosphate (MEP) and the mevalonate (MVK) pathways. In plant, MEP pathway usually working in plastids, while MVK is found in the cytosol (Rohmer 2008; Moses 2012; Abdelrahman et al. 2017). The condensation of DMAPP through a 'head-to-tail' fusion of DMAPP with one or more molecules of IPP will lead to the formation of geranyl diphosphate (GPP) with 10 carbon skeleton (C10), farnesyl diphosphate (FPP) with 15 carbon backbone (C15) or geranylgeranyl diphosphate (GGPP) with C20 structure (Oldfield and Lin 2012; Mostafa et al. 2013; Abdelrahman et al. 2017). Then, both GGPP and FPP can be fused through 'head-to-head' or 'tail-to-tail' fusions to produce different precursor molecules, including squalene and phytoene. In addition, DMAPP is also transformed by plants into isoprene units, and these isoprenoids can be cyclized to form the various terpene products. Isoprene units (C_5H_8) are the main staring building block for all terpene-related compounds through a common 'head-to-tail' fashion or 'head-to-middle' and 'head-to-head' fusions (Fig. 8.1). Thus, the terpene compounds can be classified according to the number of isoprene units they contained, where they can carry out a various function, ranging from basic cell membrane structure to specific physiological functions such as carotenoid roles in photosynthesis and quinones in electron transfer (Croteau et al. 2000; Oldfield and Lin 2012). Isoprenoids and its derived terpenes compromised the largest classes of plant natural products with approximately more than 25,000 categorized structures. For example, the terpene compounds that are consisted of two isoprene units named 'monoterpenes' with $C_{10}H_{18}$ molecular formula, contained diverse volatile compounds, including menthol, linalool, citronellol, thymol and camphor, which are commonly used for fragrance and flavor industries (Schwab et al. 2008). Similarly, the terpene compounds that are derived from three isoprene units are called 'sesquiterpenes' ($C_{15}H_{28}$) (Oldfield and Lin 2012). 'Sesquiterpenes' may be acyclic or contain rings, similar to 'monoterpenes', however, the cyclic 'sesquiterpenes' are more common compared with cyclic monoterpenes (Moses 2012). 'Sesquiterpenes' compromised different cyclic compounds such as zingiberene, caryophyllene, vetivazulene, longifolene, copaene, the alcohol patchoulol and guaiazulene (Fig. 8.1). On the other hand, the terpene class that derived from four isoprene unit fusions are called 'diterpenes' with molecular formula $C_{20}H_{32}$, and the gibberellin hormone, retinol, phytol and the plant derived anticancer taxol are members of 'diterpenes' (Croteau et al. 2000; Moses 2012). Another rare terpenes class derived from five isoprene units are called 'sesterterpenes' with the molecular formula $C_{25}H_{40}$, which have been identified in *Salvia* sp. and *Leucosceptrum* sp. medicinal plants (Choudhary et al. 2004). On the other hand, the 'triterpenes' with $C_{30}H_{48}$ molecular formula, which derived from six isoprene unit fusions are the most large class of molecules, including saponin, phytosterol and brassinosteroid-related molecules (Biswas and Dwivedi 2019).

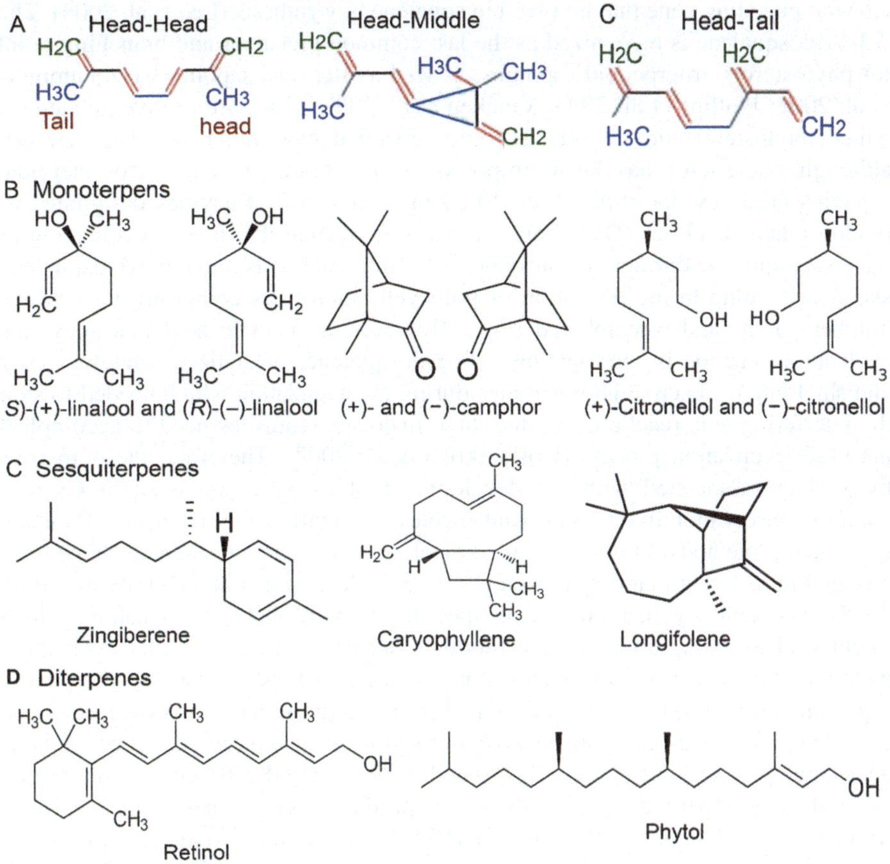

Fig. 8.1 Different linkage forms of isoprene units for terpene biosynthesis. (**a**) Heat-to-head fusion, head-to-middle fusion and head-to-tail fusions of isoprene units. (**b–d**) Mono-, sesqi- and di-terpene-related compounds, derived from different plants

Extensive efforts have been conducted in the last decades to unlock the saponin biosynthesis pathway in plants. Although the general pathways and enzymes that are intricate in saponin biosynthesis have been reported, especially for triterpene saponins, the identification of downstream regulatory genes which are responsible for the diverse array of saponin structures within different plant species is still limited (Haralampidis et al. 2002; Augustin et al. 2011). The biochemical and structure background of aglycones such as oleanolic acid and disogene are the key features for the differentiation between triterpene and steroidal saponins, respectively. Both sapogenin/aglycone types are thought to be originated from 2,3-oxidosqualene, a central molecule in phytosterol biosynthesis pathway. This hypothesis was evident by the positive correlations between saponins and sterols in transgenic ginseng (*P. ginseng*) plant overexpressing *squalene synthase*, enzyme-encoding gene that involved in squalene biosynthesis (Lee et al. 2004). The 2,3-oxidosqualene is recognized as the last common precursor and branching point for phytosterols, triterpenoid saponins, as well as steroidal saponins (Haralampidis et al. 2002; Phillips et al. 2006; Vincken et al. 2007). The further diverge steps at which phytosterol and steroidal saponin separated have not been identified yet, although cholesterol has been proposed as a potential precursor for steroidal saponins (Kalinowska et al. 2005; Vincken et al. 2007). Enzymes designated as oxidosqualene cyclases (OSCs) are known to be responsible for the cyclization of 2,3-oxidosqualene through the introduction of internal bonds into the oxidosqualene skeleton, leading to the formation of polycyclic molecules compromising various numbers of 5- and 6-membered rings. However, a catalytic acid that starts the cyclization reaction by protonating 2,3-oxidosqualene, specialized catalytic cavity and shielding of reactive intermediates during the cyclization which needed to stop the interfering side reactions, are the three major prerequisites need to accomplish the OSC cyclization process (Kolesnikova et al. 2007). Therefore, the numerous by-products associated with the development of major products of OSCs is a common observation in different plant species, especially with increasing efficiency of detection methods. In general, nine key classes of triterpene skeletons have been reported in different plants (Vincken et al. 2007; Augustin et al. 2011; Moses et al. 2014). The OSC cyclization can be specific or multifunctional in nature, which might lead to a single or multiple products derived from a particular cyclization pathway. For example, oleananes one of the most rich type of triterpenoid sapogenins found in nature are originated from β-amyrin due to 2,3-oxidosqualene cyclization by β-amyrin synthase (BAS) enzyme. In *Arabidopsis*, out of 13 *OSC* genes, two genes (*ATLUP2/At1g78960* and *AtBAS/* At1g78950) have been reported to produce β-amyrin as one of their key products, while three genes (*BARS1/ At4g15370*, *CAMS1/At1g78955* and *LUP1/At1g78970*) also produce β-amyrin as a minor by-product, indicting higher redundancy in *OSC* in Arabidopsis (Shan et al. 2008; Martelanc et al. 2007; Augustin et al. 2011). The triterpene aglycones are usually tailored by a series of cytochrome P450-dependent monooxygenases (CYP P450s), leading to an increase in structural diversity of the aglycone backbone. This CYP P450-induced oxidations at different positions of the triterpene aglycone are subsequently modified by transferase families, including malonyltransferases, UDP-dependent glycosyltransferases (UGTs), acyltransferases and methyltransferases for the addition of further functional groups (Moses et al. 2014). Similarly, the tetracyclic cycloartenol precursor, is derived from the cyclization of 2,3-oxidosqualene through cycloartenol synthase (CAS) enzyme (Abdelrahman et al. 2017). In addition, many phytosterols are thought to be derived from the cycloartenol precursor, including the C27-cholesterol, C28-campesterol and C29-sitosterol (Moses et al. 2014). The steroidal saponins are generated by a complex of oxygenation and glycosylation reactions of the cholesterol backbone to produce the furostanol or spirostanol derivatives with a fused O-heterocycle in their core aglycone structure (Augustin et al. 2011; Moses et al. 2014; Abdelrahman et al. 2017). The aglycones are tailored with oxidoreductases followed by glycosylation with multiple sugar chains (Friedman 2006; Abdelrahman et al. 2017).

8.2 Metabolic Engineering of Saponins

Plant cell and tissue cultures have been proposed as potential substitute to evade problems associated with saponin production. In vitro tissue cultures have several advantages, including high efficiency to improve and manipulate the production of desired saponin compounds, ensure high product quality and yield, and overcome germination and plant heterogeneity associated problems (Lee et al. 2018). Despite this advantages, only limited saponin compound mainly from medical plants have been produced from cell suspension cultures. For example, cell suspension culture of *Kalopanax septemlobus* woody plant, has been recently reported, by using *Friable calli* cell suspension culture (Lee et al. 2018). Results indicated that maximum amount of total saponin contents (1.56 mg 60 ml^{-1} suspension) was achieved during the initial 15 days of incubation. In addition, the total saponin production in *K. septemlobus* cell suspension was increased by addition of 1 μM coronatine (COR) into the cell suspension compared with the non-treated cells (Lee et al. 2018). This increase was highly correlated with the increase in the expression of *BSA* gene in COR-treated cell suspension compared with non-treated cell, resulting in higher contents of oleanolic acid, a key substrate for oleanane-type triterpene saponins (Lee et al. 2018). On the other hand, methyl jasmonate-treated cells showed limited improvement in the contents of total saponin levels than COR-treated cells. These results indicated that COR is an efficient elicitor for producing saponins of *K. septemlobus* cell suspension, and can be explored for other phytochemicals (Lee et al. 2018). A low-cost alternative method was also developed to generate artemisinin, a sesquiterpenoid compound from artemisinic acid using fermentation (Roth and Acton 1989; White 2008). For example, Ro et al. (2006) developed a genetically modified *A. annua* yeast strain to increase the expression of *amorphadiene synthase* (*ADS*), *CYP71AV1*, and cytochrome P450 reductase (*CPR*) to improve the production of artemisinin biosynthesis. Results indicated that, the recombinant yeast can produced a great level up to 115 mg l^{-1} artemisinic acid, a major precursor for artemisinin, and further improvements in the fermentation process can even increase the artemisinic acid up to 2.5 g l^{-1} (Ro et al. 2006; Lenihan et al. 2008). Langhansová et al. (2005) were able to develop cell suspensions cultures and root callus of *P. ginseng* for saponin production under large-scale bioreactors. In general the *P. ginseng* cells generated high yields of total saponins, in both callus, and cell suspension in bioreactor. However, total ginsenoside contents and the production of a particular ginsenosides was different between tissue cultures and cultivation systems (Langhansová et al. 2005). For example, the adventitious root cultures produce ginsenoside profile similar to native *P. ginseng* roots. However, the content of saponins in Erlenmeyer flasks was only 1.8% of dry mass, and in bioreactor was 1.5% of the dry mass, which is lower than the saponin contents in native roots (Langhansová et al. 2005). In addition, *P. ginseng* cell suspensions exhibited a different profile of specific saponins, with high induction of the ginsenosides Rb1 and Rg1 compared with the individual saponin profile in the native *P. ginseng* roots (Langhansová et al. 2005). Likewise, tissue culture of the medicinal plant *Primula veris* was established, and primula acid I, the main saponin in *P. veris* was detected (Okršlar et al. 2007). However, the average levels of primula acid I contents in suspension cell cultures and callus was eight times lower than plant gown in normal soil conditions. Although this cell culture system produced lower primula acid I saponin content, it remains as valuable alternative for the production of primula acid I, because *P. veris* is highly endangered and protected plant species (Okršlar et al. 2007). Likewise, a recombinant yeast system for investigating the functions of P450 species in glycyrrhizin biosynthesis was developed (Seki et al. 2008). *BAS* was constitutively expressed, resulting in higher accumulation β-amyrin in the yeast, then *CYP88D6* was co-expressed with a *CPR*. Results showed higher yields of 11α-hydroxy-β-amyrin and 11-oxo-β-amyrin after 2 days of incubation of the cell culture (Seki et al. 2008). Likewise, the genetic engineering of sterol biosynthesis in recombinant yeast system was established to improve the productivity of β-amyrin, and subsequently might also enhance the availability of 11-oxo-β-amyrin. In another study, Kirby et al. (2008) use recombinant yeast system expressing *Artemisia annua BAS* and truncated HMGR, 6 mg l^{-1} of β-amyrin content was obtained. In addition, considerable amount of squalene was accumulated in yeast indicating that the yeast can produce more β-amyrin, improving the incubation (Kirby et al. 2008). The variation in saponins of *Glycyrrhiza* spp. might be resulted from the diversity of UGTs and P450 involved in the downstream decoration of the saponin, and thus evaluation of the variations in these enzymes can be useful to improve their activities and increase the saponin diversity.

Several attempts have been carried out to increase the triterpene saponins in plants (Lee et al. 2004; Seo et al. 2005; Hey et al. 2006; Muñoz-Bertomeu et al. 2007; Lu et al. 2008; Kim et al. 2010). For example, the overexpression of upstream MVK biosynthesis pathway, including *3-hydroxy-3-methylgulutaryl-CoA reductase* (*HMGR*), *FPP synthase* (*FPS*), and *squalene synthase* (*SQS*) has been conducted to increase the productivity of triterpenes (Harker et al. 2003; Lee et al. 2004; Seo et al. 2005; Hey et al. 2006; Muñoz-Bertomeu et al. 2007; Lu et al. 2008; Kim et al. 2010). Although, the production of triterpenes increased relative to plant dry weight, the transgenic plants exhibited a growth inhibition phenotypes, mostly due to the metabolic imbalance (Shim et al. 2010). Therefore, further elucidation of MVK biosynthetic mechanisms is required to enhance triterpene saponin productivity in plants. In another studies. The expression of *Sad* genes in the root epidermal cells of oat plants induced the accumulation of avenacin saponin (Mylona et al. 2008; Mugford et al. 2009). On the other hand, saponin-deficient (*sad*) mutants couldn't produce a detectable amounts of avenacins, and substantially were more susceptible to *G. graminis* var. *tritici* infection than wild-type plants, providing a further link between saponin deficiency and increased disease susceptibility (Papadopoulou et al. 1999). In addition, comparative metabolomic and transcriptomic profiling also displayed good correlations between the expressions of the saponin biosynthetic genes and their accumulations, indicating that saponin production is mostly regulated at the transcription level, and also indicated that specific transcription factors for saponin biosynthesis is exist (Matsuda et al. 2010). Therefore, the engineering of saponin-related transcriptional factors could be a potential way to genetically modify the saponin biosynthetic pathway and to enhance the production by the optimizing the expression levels of multiple pathway (Hirai et al. 2007; Gonzalez et al. 2008). Saponins are known to be accumulated in vacuoles, and thus, the presence of a vacuolar transporters of saponin compounds might exist, and thus genetic engineering of saponin transporter genes can be also an efficient method to increase saponin accumulation (Mylona et al. 2008; Hayashi et al. 1996; Kurosawa et al. 2002). To date, only ATP-binding cassette transporter (NpPDR1), a terpenoid transporter that involved in secretion of an antifungal diterpene sclareol has been reported in tobacco (Jasiński et al. 2001). Therefore, by using omics technologies, it might be possible to narrow down the transcription factor(s) and transporter genes in plant secondary metabolism (Hirai et al. 2007; Sawada et al. 2009).

8.3 Conclusion

So far, all genes encoding the proteins involved in saponin biosynthetic pathway has not been achieved yet. For instance, only CYP88D6 has been reported in glycyrrhizin biosynthetic pathway which needs two P450 species and two UGTs to accomplish the production. Currently, with the development in metabolomic and transcriptomic technologies, it might be possible to accelerate the identification of plant saponin metabolisms. Integrated omics technology with of the current advanced genomic sequencing can enhance and accelerate gene discovery, and help for our understanding of the regulatory mechanisms of the expression of saponin biosynthetic genes.

References

Abdelrahman M et al (2017) RNA-sequencing-based transcriptome and biochemical analyses of steroidal saponin pathway in a complete set of *Allium fistulosum-A. cepa* monosomic addition lines. PLoS One 13:e0190813

Augustin JM, Kuzina V, Andersen SB, Bak S (2011) Molecular activities, biosynthesis and evolution of triterpenoid saponins. Phytochemistry 72:435–457

Biswas T, Dwivedi UN (2019) Plant triterpenoid saponins: biosynthesis, *in vitro* production, and pharmacological relevance. Protoplasma 256:1463–1486

Choudhary MI, Ranjit R, Rahman A-ur, Shrestha TM,Yasin A, Parvez M (2004) Leucosceptrine (I)—A novel sesterterpene with prolylendopeptidase inhibitory activity from *Leucosceptrum canum*. J Org Chem 69(8):2906–2909

Croteau R, Kutchan TM, Lewis NG (2000) Natural products (secondary metabolites). In: Buchanan B, Gruissem W, Jones R (eds) Biochemistry and molecular biology of plants. American Society of Plant Biologists, pp 1250–1318

Friedman M (2006) Potato glycoalkaloids and metabolites: roles in the plant and in the diet. J Agric Food Chem 54:8655–8681

Gonzalez A, Zhao M, Leavitt JM, Lloyd AM (2008) Regulation of the anthocyanin biosynthetic pathway by the TTG1/bHLH/Myb transcriptional complex in *Arabidopsis* seedlings. Plant J 53:814–827

Haralampidis K, Trojanowska M, Osbourn AE (2002) Biosynthesis of triterpenoid saponins in plants. Adv Biochem Eng Biotechnol 75:31–49

Harker M, Holmberg N, Clayton JC, Gibbard CL, Wallace AD, Rawlins S, Hellyer SA, Lanot A, Safford R (2003) Enhancement of seed phytosterol levels by expression of an N-terminal truncated *Hevea brasiliensis* (rubber tree) 3-hydroxy-3-methylglutaryl-CoA reductase. Plant Biotechnol J 1:113–12110

Hayashi H, Nishiyama Y, Tomizawa N, Hiraoka N, Ikeshiro Y (1996) UDP-glucuronic acid: triterpene glucuronosyltransferase activity in cultured licorice cells. Phytochemistry 42:665–666

Hey SJ, Powers SJ, Beale MH, Hawkins ND, Ward JL, Halford NG (2006) Enhanced seed phytosterol accumulation through expression of a modified HMG-CoA reductase. Plant Biotechnol J 4:219–229

Hirai MY, Sugiyama K, Sawada Y, Tohge T, Obayashi T, Suzuki A, Araki R, Sakurai N, Suzuki H, Aoki K, Goda H, Nishizawa OI, Shibata D, Saito K (2007) Omics-based identification of *Arabidopsis* Myb transcription factors regulating aliphatic glucosinolate biosynthesis. Proc Natl Acad Sci USA 104:6478–6483

Jasiński M, Stukkens Y, Degand H, Purnelle B, Marchand-Brynaert J, Boutry M (2001) A plant plasma membrane ATP binding cassette-type transporter is involved in antifungal terpenoid secretion. Plant Cell 13:1095–1107

Kalinowska M, Zimowski J, Pączkowski C, Wojciechowski ZA (2005) The formation of sugar chains in triterpenoid saponins and glycoalkaloids. Phytochem Rev 4:237–257

Kim OT, Kim SH, Ohyama K, Muranaka T, Choi YE, Lee HY, Kim MY, Hwang B (2010) Upregulation of phytosterol and triterpene biosynthesis in *Centella asiatica* hairy roots overexpressed ginseng farnesyl diphosphate synthase. Plant Cell Rep 29:403–411

Kirby J, Romanini DW, Paradise EM, Keasling JD (2008) Engineering triterpene production in *Saccharomyces cerevisiae* β-amyrin synthase from *Artemisia annua*. FEBS J 275:1852–1859

Kolesnikova MD, Obermeyer AC, Wilson WK, Lynch DA, Xiong Q, Matsuda SPT (2007) Stereochemistry of water addition in triterpene synthesis: the structure of arabidiol. Org Lett 9:2183–2186

Kurosawa Y, Takahara H, Shiraiwa M (2002) UDP-glucuronic acid:soyasapogenol glucuronosyltransferase involved in saponin biosynthesis in germinating soybean seeds. Planta 215:620–629

Langhansová L, Marsik P, Vanek T (2005) Production of saponins from *Panax ginseng* suspension and adventitious root cultures. Biologia Plantarum 49:463–465

Lee MH, Jeong JH, Seo JW, Shin CG, Kim YS, In JG, Yang DC, Yi JS, Choi YE (2004) Enhanced triterpene and phytosterol biosynthesis in Panax ginseng overexpressing squalene synthase gene. Plant Cell Physiol 45:976–984

Lee JK, Eom SH, Hyun TK (2018) Enhanced biosynthesis of saponins by coronatine in cell suspension culture of *Kalopanax septemlobus*. 3 Biotech 8:59

Lenihan JR, Tsuruta H, Diola D, Renninger NS, Regentin R (2008) Developing an industrial artemisinic acid fermentation process to support the cost-effective production of antimalarial artemisinin-based combination therapies. Biotechnol Prog 24:1026–1032

Lu HY, Liu JM, Zhang HC, Yin T, Gao SL (2008) Ri-mediated transformation of *Glycyrrhiza uralensis* with a *squalene synthase* gene (*GuSQS1*) for production of glycyrrhizin. Plant Mol Biol Rep 26:1–11

Martelanc M, Vovk I, Simonovska B (2007) Determination of three major triterpenoids in epicuticular wax of cabbage (*Brassica oleracea* L.) by high-performance liquid chromatography with UV and mass spectrometric detection. J Chromatogr A 1164:145–152

Matsuda F, Hirai MY, Sasaki E, Akiyama K, Yonekura-Sakakibara K, Provart NJ, Sakurai T, Shimada Y, Saito K (2010) AtMetExpress development: a phytochemical atlas of *Arabidopsis thaliana* development. Plant Physiol 152:566–578

Moses T (2012) Biology Metabolic engineering for production of triterpenoid saponin building blocks in plants and yeast. PhD thesis

Moses T, Papadopoulou KK, Osbourn A (2014) Metabolic and functional diversity of saponins, biosynthetic intermediates and semi-synthetic derivatives. Crit Rev Biochem Mol Biol 49:439–462

Mostafa A et al (2013) Aginoside saponin, a potent antifungal compound, and secondary metabolite analyses from *Allium nigrum* L. Phytochem Lett 6:274–280

Mugford ST, Qi X, Bakht S, Hill L, Wegel E, Hughes RK, Papadopoulou K, Melton R, Philo M, Sainsbury F, Lomonossoff GP, Roy AD, Goss RJ, Osbourn A (2009) A serine carboxypeptidase-like acyltransferase is required for synthesis of antimicrobial compounds and disease resistance in oats. Plant Cell 21:2473–2484

Muñoz-Bertomeu J, Sales E, Ros R, Arrillaga I, Segura J (2007) Up-regulation of an N-terminal truncated 3-hydroxy-3-methylglutaryl CoA reductase enhances production of essential oils and sterols in transgenic *Lavandula latifolia*. Plant Biotechnol J 5:746–758

Mylona P, Owatworakit A, Papadopoulou K, Jenner H, Qin B, Findlay K, Hill L, Qi X, Bakht S, Melton R, Osbourn A (2008) Sad3 and Sad4 are required for saponin biosynthesis and root development in oat. Plant Cell 20:201–212

Okršlar V, Plaper I, Kovač M, Erjavec A, Obermajer T, Rebec A, Ravnikar M, Žel J (2007) Saponins in tissue culture of *Primula veris* L. In Vitro Cell Dev Biol Plant 43:644–651

Oldfield E, Lin F-Y (2012) Terpene biosynthesis: modularity rules. Angew Chem Int Ed Engl 51:1124–1137

Paek KY, Murthy HN, Hahn EJ, Zhong JJ (2009) Biotechnology in China I. 113:151–176

Papadopoulou K, Melton RE, Leggett M, Daniels MJ, Osbourn AE (1999) Compromised disease resistance in saponin-deficient plants. Proc Natl Acad Sci U S A 96:12923–12928

Phillips DR, Rasbery JM, Bartel B, Matsuda SPT (2006) Biosynthetic diversity in plant triterpene cyclization. Curr Opin Plant Biol 9:305–314

Ro DK, Paradise EM, Ouellet M, Fisher KJ, Newman KL, Ndungu JM, Ho KA, Eachus RA, Ham TS, Kirby J, Chang MC, Withers ST, Shiba Y, Sarpong R, Keasling JD (2006) Production of the antimalarial drug precursor artemisinic acid in engineered yeast. Nature 440:940–943

Rohmer M (2008) From molecular fossils of bacterial hopanoids to the formation of isoprene units: discovery and elucidation of the methylerythritol phosphate pathway. Lipids 43:1095–1107

Roth RJ, Acton N (1989) A simple conversion of artemisinic acid into artemisinin. J Nat Prod 52:1183–1185

Sawada Y, Toyooka K, Kuwahara A, Sakata A, Nagano M, Saito K, Hirai MY (2009) *Arabidopsis* bile acid:sodium symporter family protein 5 is involved in methionine-derived glucosinolate biosynthesis. Plant Cell Physiol 50:1579–1586

Schwab W, Davidovich-Rikanati R, Lewinsohn E (2008) Biosynthesis of plant-derived flavor compounds. Plant J 54:712–732

Seki H, Ohyama K, Sawai S, Mizutani M, Ohnishi T, Sudo H, Akashi T, Aoki T, Saito K, Muranaka T (2008) Licorice β-amyrin 11-oxidase, a cytochrome P450 with a key role in the biosynthesis of the triterpene sweetener glycyrrhizin. Proc Natl Acad Sci USA 105:14204–14209

Seo JW, Jeong JH, Shin CG, Lo SC, Han SS, Yu KW, Harada E, Han JY, Choi YE (2005) Overexpression of squalene synthase in *Eleutherococcus senticosus* increases phytosterol and triterpene accumulation. Phytochemistry 66:869–877

Shan H, Wilson WK, Phillips DR, Bartel B, Matsuda SPT (2008) Trinorlupeol: a major nonsterol triterpenoid in *Arabidopsis*. Org Lett 10:1897–1900

Shim JS, Lee OR, Kim YJ, Lee JH, Kim JH, Jung DY, Lee BS, Yang DC (2010) Overexpression of *PgSQS1* increases ginsenoside production and negatively affects ginseng growth rate in *Panax ginseng*. J Ginseng Res 34:98–103

Vincken J-P, Heng L, de Groot A Gruppen H (2007) Saponins, classification and occurrence in the plant kingdom. Phytochemistry 68:275–297

White NJ (2008) Qinghaosu (artemisinin): the price of success. Science 320:330–334

The manufacturer's authorised representative in the EU is Springer Nature Customer Service Centre GmbH, Europaplatz 3, 69115 Heidelberg, Germany. If you have any concerns regarding our products, please contact ProductSafety@springernature.com

Printed and bound by CPI Group (UK) Ltd, Croydon, CR0 4YY
28/09/2026
02224086-0015